CUT THE
CORD

CUT THE CORD

How to Achieve

Energy Independence

by Joining the

Solar-Powered

Microgrid Revolution

VINCENT BATTAGLIA, MBA

Advantage®

Cut the Cord is A Mycrogrid Electric Corp. Publication

Published by Advantage, Charleston, South Carolina.
Member of Advantage Media Group.

ADVANTAGE is a registered trademark and the Advantage colophon is a trademark of Advantage Media Group, Inc.

Printed in the United States of America.

ISBN: 978-1-59932-573-6
LCCN: 2015950220

Book design by Megan Elger.

This publication is designed to provide accurate and authoritative information in regard to the subject matter covered. It is sold with the understanding that the publisher is not engaged in rendering legal, accounting, or other professional services. If legal advice or other expert assistance is required, the services of a competent professional person should be sought.

Advantage Media Group is proud to be a part of the Tree Neutral® program. Tree Neutral offsets the number of trees consumed in the production and printing of this book by taking proactive steps such as planting trees in direct proportion to the number of trees used to print books. To learn more about Tree Neutral, please visit **www.treeneutral.com**. To learn more about Advantage's commitment to being a responsible steward of the environment, please visit **www.advantagefamily.com/green**

Advantage Media Group is a publisher of business, self-improvement, and professional development books and online learning. We help entrepreneurs, business leaders, and professionals share their Stories, Passion, and Knowledge to help others Learn & Grow. Do you have a manuscript or book idea that you would like us to consider for publishing? Please visit **advantagefamily.com** or call **1.866.775.1696**.

This publication could not have been brought to light without the support, humor and energy of my wordsmith and muse Lea Goodsell. My respect and gratitude for you go beyond words Lea! And to Jason Rowe whose knowledge of my knowledge and passion to always live in "right" made the evolution of this scroll a dream come true for me... his talent and word-spice took this project from a bland-rant to a grand-stand...thank you Jason eternally! And to my father Vincent for always enthusiastically living his life fighting the good fight...with every day that passes his example reminds me, "Ho combattuto la buona battaglia, ho terminato la mia corsa, ho conservato la fede."

TABLE OF CONTENTS

DEDICATION

My name is Vincent Battaglia, and my family name means "to battle, to fight." I've found that name perfectly suited to the industry that my forefathers and I dedicated our lives to—one of my Italian forefathers was a solar pioneer. Incidentally, my first name also means, "to win, defeat, and vanquish," which is something I fully intend to do for the benefit of the world.

This book, this labor of love (or you may call it a battle cry) is dedicated to you—yes, to you. The fact that you are reading this book deserves recognition for your step in the right direction.

I also wish to dedicate this book to several groups of people I feel very passionate about. The first of these is others like you and me who have gone before us and blazed a trail; stared down adversity; overcame challenges; and who insisted on a better, sustainable, and green-saving alternative (money and the planet) to the "umbilical cord" of the utility companies and the "power mafia" that profit from them.

The other groups I dedicate this book to are the power mafia leadership, those lobbying for them in the government, and those "wagging the dog" for them through the media. If it wasn't for the exploitation of the populace by these powers, which are empowered by lined pockets and supported by a tireless propaganda crusade against alternative energy pioneers and citizens who are brave enough to take control of their energy bill, I may never have been inspired to write this book. To each of you in these latter groups, I say: You and I both know that the sands in your monopoly's hourglass are running out. If you have a conscience, if you care about people, if you're tired of spouting desperate rhetoric and would prefer to subscribe to common sense, then by all means—join us! If not, if you prefer to continue on your path of greed, then get out of the way.

UTILITY MALPRACTICE

I ask that you bear with me for the next few pages while I make clear, in no uncertain terms, how I feel about the condition of the utility-controlled energy industry, about the corrupt individuals controlling it, and most importantly, about the very bright future that is dawning. This may come off a bit harsh at first, but if you'll bear with me I promise that by the time you finish this book, you will understand why microgrids represent the next solar renaissance, what's in it for all of us, and how, together, we can take ownership of our energy production and consumption like never before

So please indulge me for a moment while I get a few things off of my chest.

With the declaration that we were capable of identifying personal needs, the solar-plus-storage industry signaled the end of the "I" generation. Thus began the "My" generation in which we are just beginning to fulfill those needs—independent of large corporate intervention or attachment. Solar-plus-storage delivered the genesis of true independence and personalizes the experience of that independence.

There's a lot to cover in this book. So how do you eat an elephant? One bite at a time.

DEMAND MANIFESTO ...
AN EXERCISE IN UTILITY MALPRACTICE

My home state of California has a water problem. Actually, it has a water utility problem. The water utilities appear to have no concept of a tomorrow and the demands/shortfalls that will come with it. Soon Californians will have to cross over to Arizona for a glass of water because our water utilities are poorly staffed and underserve the populace with a poor capacity to plan. California's water and electric utilities are a commodity mafia. Why do we believe that the electric utility companies will deliver a sustainable model for the future when their water utility brethren have proven that they cannot?

We are going to certainly run into an electricity crisis. Microgrid technology is not only the right approach ... it is a need for our civilized society.

Experiences are changing every day on our planet.

Why should we embrace cutting the cord on cable TV companies (as an example of progress) and not on electric companies? We walk from room to room and car to office, carrying our TV monitors without cords and cables ... why shouldn't energy experiences exist just as sure and certain?

As I see it, we have no restraints at this point—technological or financial. My mentors have taught me that nothing can stop an idea whose time has come and that getting everyone onboard with the essence of the idea ultimately makes the idea a reality. So all aboard!

After living for seven years in Russia, I despise beurocracy to the point that I don't even bother to spell it correctly! I spent nearly every day of my life there tearing down central government control to expose the natural roots of the free market.

I returned to the United States only to realize that we are a society controlled by the iron fist of energy fiefdoms—oligarchs given control of the very fuel of our economic engine. When everyone had one light bulb and one television in their home, that model worked—to a point. But as we evolve, so must the patterns of our utilities.

HEADS UP UTILITY COMPANIES

To the utility companies, I say this: You were granted a license by society to establish networks (grids) in territories so that you could provide electricity and manage, grow, advance, and maintain your allotted grid section using portions of the monthly fees paid by those living in your territories and benefiting from the product (current) that you were licensed to produce. You did that. And you've profited very well. Advances in electric current delivery are emerging at a scale that is less intrusive, simpler in design, more mobile in distribution, cleaner, quieter, and at a lower cost.

Solar. Heard of it? Haven't heard of it? Well, you've had the opportunity to develop that alternative energy before the free market made it a mainstream part of the real and relevant energy mix in this country. You lost that advantage. Now, move over.

How do I know this? Because your struggle to criticize the alternative since 2006, when I installed Renova's very first solar-generating power plant using every ploy possible, resulted in my business doubling year-over-year. The scare tactics that you relied on—calling the code-compliant methods that we rigorously followed "unsafe" and "amateur"—didn't work. Our methods resulted in zero deaths, zero fires, and zero catastrophes. I call your efforts the "Exercise of Utility Malpractice." I lived with those very same tactics in Russia during the 1990s under a conquered regime shamefully pulling every string it controlled until there were few left.

Your strings are breaking as well—your employees are nearing retirement, and your brand is tarnished more every month as your end customer receives an ever-increasing bill.

This manifesto is a declaration more to point at your demise than to celebrate its own emergence as a logical and intelligent replacement whose time has come. It's sad that we do not celebrate the coming of hope in a cleaner, less restrictive, low-cost, and worry-free transformation for future generations. Once you admit your imminent fate, then proper policy can be established to allow for your departure and the emergence of the much more beneficial technology provided dependably by the sun and its solar energy. This is not about you any longer.

I am solar. Everyone reading these words is solar. Solar, as is any emerging technology when it becomes better than yesterday's standard, should not bear the brunt of finding a pasture for the irrelevant has-been it is justly replacing to die in. It is no longer our responsibility to care for and feed you or to integrate us into your prolonged betterment. It is incumbent on those policymakers (who are paying attention) to mandate your integration of me into your monopolized sphere as you wither away into dirty, mind-numbingly environment-damaging, over-priced, and technologically antiquated obscurity.

Now let's get to the good news and bright ideas.

INTRODUCTION

By now I hope your research and powers of observation have led you to the right conclusion: solar is not a fad. The sun has been around longer than all of us; has been providing energy for humankind since the dawn of man; and has been well established for some time now as a better, brighter, and (more than ever before) affordable alternative to utility grid reliance. Solar represents progress, efficiency, ingenuity, and freedom. In fact, I believe solar represents the independent spirit of America as much, if not more, than many of the progressive advances we rely on today.

The book you're reading is a call to *change* for all those who have grown up with fossil-based energy and a call to *action* for millennials who may have only spent a couple of decades on this spinning rock we call home but are old enough to know that most of the old ways of doing things are outdated, surpassed by the technology they have grown up with, and that these advances have opened doors to almost limitless possibilities for progress. Sure, I'll be sharing technical insights into microgrids and solar at the risk of revealing my inner geek—but I didn't write this book for a lecture or a TED talk. Well, actually, I *would* welcome that opportunity.

I wrote this to have a conversation with my peers who are still drinking the wrong Kool-Aid and *especially* with young families and professionals to let them know how close they are to being able to ditch the outdated concept of "leasing" utility-produced electricity for the very real and present opportunity to own it. You may think you know solar—but the microgrid is changing and will continue to change everything.

ORIGINS

For starters, I'd like to share how I arrived at the point of writing this book. As I'll reference throughout these pages, I spent a good stretch of time in post-communist Russia in the 1990s. I went there seeking fame, fortune, and world domination. All kidding aside, my colorful experiences were filled with television gladiators, brushes with the Russian mafia, Hollywood executives, and even a brief stint in acting that nearly had me carrying a sword and shield in the film *Gladiator*. I could literally write a book based on my experiences there alone (and may at some point), but what's relevant about that time in my life is that my early interests in solar were rekindled, and my distaste for a grid structure of energy delivery was born from the corruption, inefficiencies, pollution, and waste that I witnessed with Russia's centralized grid system.

Hot water pipes in Russia illustrate an extreme case of outdated centralized delivery.

I returned to the States still set on becoming an entrepreneurial success story, but it wasn't long before my sights turned to applying my gumption and competitive nature to the pursuit of making a real difference in the world around me as both a businessman and advocate for positive change. My mission led me to advance my education through graduate studies. I started Renova Solar in 2006 as an offshoot of my MBA thesis paper at the University of California, Riverside.

A year later, Renova Energy Corporation was born in the city of Palm Desert, just down Interstate 10 from Palm Springs and only a few minutes from renowned attractions like PGA West, Coachella Valley Music and Arts Festival, and Indian Wells Tennis Garden, home of the PNB Paribas Open.

Renova has grown fast from its humble-but-determined beginnings and today offers a completely integrated solar operation—from design and installation and finance through to service and the installation of advanced energy systems through RenovaPLUS Mycrogrid®. We also lead educational and training efforts in sustainable technology opportunities through the Renova Energy Academy, the region's only renewable energy, private enterprise educational program, conducted in association with top trainers regionally and nationally and through an affiliation with the University of California, Riverside's extension program.

We did all of this from our corporate offices, appropriately called "The GreenZone." That banner is now being carried at the local community college and has even made its way to a couple of area high schools, which have established energy academies focusing on renewable energy.

GreenZone, near Palm Springs, is the corporate headquarters for Mycrogrid and Renova Solar.

I founded The GreenZone to educate and empower start-up companies to join and grow the green economy, as well as to showcase the latest solar, renewable energy, and green technology, in real time, as they became available for commercial and residential buildings. In short, The GreenZone was founded and continues to serve as an incubator, education facility, showroom, and a "lighthouse" to guide and inspire local business and homeowners.

Our tireless efforts have not gone unnoticed. I'm very proud that in 2012, Renova Solar became "America's first professionally accredited solar company" by the North American Board of Certified Energy Practitioners (NABCEP), the leading authority and most-respected certification organization for those working in the solar industry. Previously, systems, panels, and components were validated after receiving NABCEP certification. Renova Solar was the first solar company in NABCEP history to receive the organization's seal of approval.

Since founding Renova Solar, my existence has been driven by education, advocacy, networking, and innovation. Like my

Battaglia family members before me, I've never been satisfied with good enough; I've always challenged the status quo and had my eyes fixed on the horizon to where solar technology was heading with the greater good in mind. My pursuit of brighter and better solar solutions allowed me to learn and experience enough to know that the future of energy distribution would be tied to storage of that energy. As long as utility companies held the stored power, cutting the cord would never be an option.

The solution to true independence would depend on delivering reliable, scalable, and affordable storage solutions to businesses and residential consumers. All other pieces for solar independence were already in place and proven, especially with zero-down lease options that make solar accessible to virtually everyone, no matter where they live.

Driven by an obsession to deliver the solution, I founded Mycrogrid in 2014 and joined forces with advanced energy management companies to offer smart-battery systems that would empower consumers to augment their energy savings and ultimately take ownership of their own private grid—within the grid—not miles away from the nearest transformer. I'll share more on our progress later on, but first, please allow me to introduce you to the microgrid if you haven't met already. If you have, then this reintroduction will be a good refresher that will provide you with additional insight to what we are calling, "The next solar renaissance."

INTRO TO MICROGRIDS

A microgrid is a stand-alone, stored energy device powered by a renewable energy source—the best, most dependable of which are solar photovoltaics (PV). They can be large in size for industrial electric load requirements or small in size when connected to a home or business customer still tied into the aging utility grid. Today, the primary purpose of that grid-tied connection is for resiliency and certainty as the utility grid begins to wind down in capability and increase in delivered cost. The rapidly evolving purpose of the near tomorrow is to cut the cord of the utility grid for stand-alone, dependable, and individual energy independence.

Microgrids use battery storage to convert the passive electric generation of solar into integrated energy storage. When adding energy storage to create a microgrid, we are ensuring the electrical system can act autonomously and uninterrupted, providing reliable power all day and all night.

A microgrid's general effect on ratepayers connected to the dying utility grid is to

1. keep electrical stress off the weak points in the grid,

2. provide instant response to demand,

3. ensure no inconvenience to ratepayers,

4. be safe and reliable, and

5. encourage the utilities to take their proper role as grid services providers and not electricity producers.

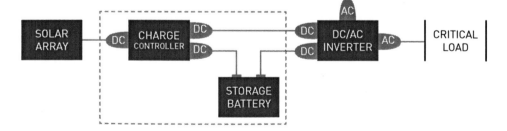

At this time, utility grid-tied microgrids allow the solar power plant to function by meeting your home's local energy demand with its clean current flow and sending excess power to the utility grid (while taking advantage of the benefit of drawing from the grid when needed). Society paid for that utility grid with trillions of dollars over the past century, so there's no reason not to remain attached until microgrid systems become second nature and maturing storage propagates the smart energy age.

Oftentimes, our solar customers are surprised to find their home without power when the utility grid blacks out, even during the day when the sun is shining. As of the writing of this book in the summer of 2015, when most solar electric systems are installed, they are tied into the utility grid without energy storage and stand-alone (islanding) capability. For safety, these solar systems automatically shut down when the utility grid shuts down so that the energy created by them is not sent down the grid, putting utility workers at risk of shock while they are trying to patch the aging utility grid back together.

Storage is beneficial to all who consume electric power—even if you haven't "gone solar" yet. Even without solar, your batteries are filled by simply connecting to the utility's macrogrid and drawing from that source. A macrogrid is simply another term for a utility's grid, as opposed to a home system with storage. Going forward, we will simply refer to the utility macrogrid as the grid. Battery storage allows solar and non-solar users to shift load and store energy when utility grid power is less costly or the sun is shining on your solar panels, so that you can let that power loose during the times when the utility grid charges too much. With solar and battery storage—you're a microgrid, but even with battery storage alone, you're still adopting better energy-consuming practices. Battery storage can also serve as a backup should the grid go down. Battery storage has evolved to more than the familiar car battery that maintains a firm charge at all times. Today's batteries can tap full potential, standing at the ready to charge again and again to their depths. Lithium-ion batteries as storage devices are low-maintenance, need little ventilation, are lighter weight, and perform better in hot and cold temperatures. They also have a longer lifespan than their predecessors. Adding solar to storage closes the loop, forms your microgrid, and opens an energy-independent future for you and yours.

With batteries as mobile and simple storage devices, storage of energy is signaling the last days of solar photovoltaics being the final solution to the consumer.

> Once upon a time, utility monopolization of energy provision made good sense. By giving complete control to utilities (who were in turn independently responsible for providing power to all who lived in their designated service areas), synergy in service was achieved; rates dropped in the face of rising demand; and ample fossil fuel resources ensured that the populace would have warm homes, hot food, cold drinks, and electrically powered entertainment until kingdom come.
>
> Today our perpetually powered society has outgrown the grid, forcing utilities to turn to solar and other alternative energy to keep pace with demand. It seems obvious to me, but you have to ask yourself, "If solar is such a bad idea, poses an eminent threat to grid-connected consumers everywhere, and is a cost-prohibitive alternative being irresponsibly promoted by opportunis-

> tic solar contractors (as utility company talking heads and media slur campaigns assert), why are utility companies building massive solar installations and lobbying for control of rooftop solar while fighting to keep much of the surplus energy generated by private solar users in exchange for little or no cost to themselves?"

Harnessing the power of the sun dates back to the discovery of fire and is even featured in mythology where the likes of Archimedes is said to have saved ancient Greece by using the sun to create a laser beam that sunk a fleet of ships manned by Roman invaders—A&E's *MythBusters* and other scholars have actually confirmed it's possible.

My passion for solar goes a long way back. You see, solar is a family trade dating back to my ancestor, Alessandro Battaglia, who filed for the first solar collector patent in 1886. I'll share more of his story later in this book, but Alessandro is known as a pioneer in the development of concentrating solar power (CSP) using flat or nearly flat reflectors. His invention was his answer to the challenges of building large collectors. He also pioneered a new way to store that energy with boilers, creating one of the earliest microgrids. Not much else is known about Alessandro Battaglia, but this historical record (filed at a time when the foundations of modern solar technology were being laid) demonstrates that the Battaglia family has a long history of keeping a finger on the pulse of solar energy trends and in pioneering new solar technology.

As a young man in 1992, I traveled to Europe, Eastern Europe, and, ultimately, Russia with visions of making my mark. I believed I could be anything and everything to anyone and everyone by testing myself and accomplishing something original in a post-communist country that was just as determined to find its identity and place in the world as I was.

My Russian backstory aside, while there I personally observed three things that played a major role in where and who I am today:

1. The incredible waste that exists in regard to energy, the environment, and governance.

2. Just how far money and power will allow those in possession to manipulate, coerce, and control the populace.

3. My first sighting of solar used practically as an alternative electricity generating source.

This third observation was only a glimpse. It happened on a winter's night in the Latvian capital of Riga in 1997. I didn't recognize the application at first, but the sight of a rooftop photovoltaic (PV) solar system alongside a solar thermal system on an apartment building in the center of town had my head buzzing.

The site of that solar system planted a seed that would ultimately come to fruition nearly a decade later under the shelter of, and inspired by, the freedom we enjoy (and too often take for granted) here in the United States of America.

Freedom.

The word conjures a multitude of images, shapes ideologies, and inspires action. Everyone chases it in one form or another regardless of age, gender, creed, or color. Those in bondage dream of it. Patriots fight for and defend it, but what does freedom mean to you? A quick search on Merriam-Webster's website will show you that freedom is defined as:

✓ The absence of necessity, coercion, or constraint in choice or action.

✓ Liberation from slavery or restraint or from the power of another.

✓ Boldness of conception or execution.

✓ Unrestricted use.

Sounds good, doesn't it?

No one wants to be forced into making a choice or taking an action. No one likes to be subject to someone more powerful, and those at the mercy of others would never have chosen that life. No one prefers to be burdened. No one aspires to be timid. No one gets giddy over restrictions, right?

Yet we allow ourselves to become or to be controlled by each of those things in various ways every day! Think about it. Have you ever worked a job that you hated for painfully low wages instead of pursuing your passion and purpose in life? How many of those hard-earned and scarce dollars have you never seen because of taxes? The list goes on, but you see my point, right?

So what does freedom actually free you from?

✓ Necessity, coercion, or constraint in choice or action

✓ Power of another

✓ Something onerous

Okay, I admit it. I had to look that last one up. Onerous is synonymous with burdensome, difficult, hard, severe, oppressive, challenging, tiring, taxing, and demanding. Call me crazy, but I'd say that's a fitting description of what it feels like to pay a summer electric bill anywhere that air conditioning is a necessity!

On the other hand, renewable energies like solar power are defined by traits that actually make one free by definition:

✓ Boldness of conception or execution

✓ Unrestricted use

When you look at it that way, renewable energy is downright American! In fact, I believe current and future solar technology represents the very essence and vision of independence that America was founded on. (It may even hold the secrets to true bipartisanship with its fiscally conservative, yet environment-friendly nature.)

Clean energy isn't just a good idea or a passing fad. It's been around longer than anyone at any time, and it is here to stay. It's healthier and saves you money—so it can help you to live not only a better life today but possibly a longer one.

In case you haven't taken a look at solar technology lately, it's bold in conception and execution and has become more accessible, affordable, and effective with each new sunrise. Today, people from all walks have access to the endless power supply that the sun provides on a daily basis. But hang on—it gets better.

You already know that solar can save you money on your electric bill, but what if I told you that you could own your own personal power grid ... a grid within the grid? What if I told you that this personal power grid could guard your home against power outages, power your personal office, or keep your ice cream cold without *any* connection to the power company? What if I told you that this same technology would soon be able to power an entire house with a tablet-controlled power unit no bigger than a file cabinet?

It sounds like science fiction, I know—but it's here.

Before I go any further, I want to acknowledge the hard work, ingenuity, and success of both solar pioneers and (exceptionally) intuitive utility leadership who have successfully designed, implemented, and blazed a trail for the microgrid journey toward mass assimilation.

All rants and calls to action aside, the purpose of this book is *not* to claim that mine is the first or lone voice regarding the microgrid. There are many historical and current examples of successful microgrid operations, often driven by necessity in the absence of a grid-tied alternative.

As I'll touch on later in this book, solar-powered microgrids are already replacing diesel generator-powered villages and health facilities in developing countries such as Africa, India, China, as well as in remote areas of Eastern Europe and similar off-grid locations where there is no access to power otherwise and where lives depend on access to electricity.

There are many great examples of successful microgrid implementation within our borders here in the US as well. Historic landmarks such as Alcatraz Island and many national parks are relying on microgrid power at this very moment.

The United States military uses microgrids powered almost entirely by renewable energy through its SPIDERS program (Smart Power Infrastructure Demonstration for Energy Reliability and Security).

Visionary professors and millennial college students on campuses in Delaware, Wisconsin, Washington, Massachusetts, California, and beyond are not only using microgrids on campus but are developing the next, big breakthroughs in solar, battery storage, and microgrid technologies.

And yes, as I'll doggedly point out in the following pages, even the utility companies are leveraging solar and microgrid technology to enhance their outdated and overtaxed grid delivery system, sustaining energy for remote customers who would otherwise wait unbearable periods of time for restoration of service when the grid inevitably fails, seeking to own residential solar and microgrid technology, and cunningly shot gunning press releases on their noble vision for a "smart grid" that is more responsive, dependable, sustainable, and customizable with renewable energy-powered microgrids at the heart of it all.

It's not an original idea (as I'll explain—insert sarcasm here), but it's nice of them to consider joining the party, *even if it is at your expense.*

What this book *is* intended to suggest and demonstrate is that I, along with the brilliant minds I've surrounded myself with, have put the "my" into microgrid.

I offer much deserved congratulations, thanks, and respect to all who have embraced the independence-enabling power of the microgrid. What I'm about to share with you is how we are making it accessible for you today right where you live and how microgrids will give you control over your own power supply and usage, ultimately, with complete independence from the utility grid if you so choose, even if there are power lines overhead.

SOLAR FACTOID

Did you know that in one minute the sun produces enough energy to power the entire planet's electric needs for a full year?

CHAPTER ONE

REASONABLE, NOT RADICAL

DISPELLING THE STEREOTYPE OF LIVING "OFF THE GRID"

Have you ever heard someone talking about "going off grid?" What was your reaction? In my experience, when someone hears the phrase "living off grid," one of two reactions typically follow: either their brow furrows a bit with a look of confusion, or their brow furrows a bit with a look of distrust and suspicion.

The first reaction is common for those who have never really engaged in conversation or conducted personal research on the topic of solar energy. I like chatting with these folks because they haven't heard me preach the message of renewable, inexhaustible energy a hundred times before (like my long-suffering family and friends have). And they're far less likely to roll their eyes when I launch into "solar evangelist" mode and give a passionate solar sermon filled with technical terms like photovoltaic, kilowatts, inverters, and arrays.

The second reaction is often the byproduct of stereotypes and misperceptions. It may come with a mental image of hairy, anti-establishment types who bear arms (lots of them) and plot anarchy from some remote cabin. I enjoy chatting with these types because it means I have the opportunity to change paradigms and paint a realistic and relevant picture of what gaining energy independence looks like.

Historically, living "off grid" simply meant that an individual either couldn't depend on the power company due to power access issues, or they chose not to rely on public utilities for simplicity and cost-savings sake.

Sounds reasonable enough, doesn't it? Choosing to leave the hustle and bustle behind and retreat to a quiet life surrounded by nature is a perfectly rational choice to make. In fact, a lot of work-weary people pay good money for only a few days of vacation in such a setting! Those choosing to trade one lifestyle for another have taken the next logical step. Why is it that these logical individuals are often equated with a few misguided radicals seen on the evening news or featured in documentary films?

Let's pose a rhetorical question to find the answer: If you hold an obscenely lucrative share of a global monopoly that entire nations are dependent on every day, and then a more dependable, inexpensive, and inexhaustible alternative is presented to the world that threatens to render your monopoly (and your fortunes) obsolete, what lengths do you suppose you'll go to in order to keep your profits coming in for as long as possible?

If the competitive alternative to your monopoly has facts, history, and common sense on its side, then your best bet is to throw your weight and money into raising doubt, creating fear, and lobbying the powers that be to aid you in delaying the inevitable, right? If you doubt it, consider historical precedence. After reading the following short list of words, take a minute and think about the widely reported scandals, controversies, and subsequent efforts to avoid responsibility or conceal the truth regarding each of these provocative topics:

» oil

» pharmaceuticals

» tobacco

» political corruption

» war

» terrorism

The truth is that those who make their fortunes or wield power at the expense of others are never willing to let go of that power and the perks that go with it—at least not without a fight. If we can agree that's true, then consider one more addition to the list: *electric utility companies.*

I know. We generally don't think of utilities as inherently evil and worthy of our distrust the way we often do others—such as politicians, defense attorneys, insurance salespeople, and telemarketers. Sure, we'll complain about rate hikes and rolling blackouts, but we tend to handle these hardships as rant opportunities rather than signs that change is needed—much like we do at the gas pump. Did I mention oil?

True, there's nothing shady about electricity, but if supply of a vital resource is perpetually outpaced by demand—and rather than tackle the challenge with a readily available solution, you choose instead to leverage the shortage into higher profits—that's bad-form capitalism. It's also known as exploitation.

I'm a business owner striving for profitability. I wholeheartedly support anyone chasing the American dream. Design the best fashion. Create the best art. Make the best gourmet food. Produce a better vehicle. Invent a better mousetrap. If you accomplish any of those things, I believe that you deserve to charge what you're worth. Supply and demand is not the issue. Claiming ownership of a resource that belongs to and is relied on by everyone, and then using your role as a provider to hold the public hostage to price gouging is not fair trade, and that's the issue.

Electric utilities have much to lose when consumers finally understand that dependence on the grid is already a dated concept and will be unnecessary before we know it. That's why utilities use fear tactics to delay the inevitable. With the help of the media, utility companies have painted an erroneous picture of the typical alternative energy producer/consumer as being a "freeloader" who "represents a real threat" to the utility company's ability to serve the rest of society with "reliable" power. In reality, the typical solar/renewable energy user probably looks a lot like you or someone you know.

In a recent study, the Center for American Progress (CAP) reported that in Arizona, California, and New Jersey—the three largest solar markets in the United States—the majority of solar panels are being installed in homes with median incomes ranging from $40,000 to $90,000. CAP also found that the emerging solar markets in Maryland, Massachusetts, and New York are following these figures closely, with more than 80 percent of the installations in New York also occurring in areas with a median income of $40,000–$90,000. I don't know about you, but that sounds like middle Americans to me with nothing fringe, nefarious, or subversive about them.

The irony of these smear tactics is that utility companies are becoming increasingly dependent on renewable energy to generate power for the masses. Great examples are the fields of turbines located just outside of my home base in Palm Desert, California (as seen in the early minutes of *Mission Impossible 3*, during the helicopter chase over, under, and through the massive, spinning turbine blades).

The image of vast fields of massive turbines generating renewable energy is something to behold, for sure. And if one didn't know any better, it would sound like a logical, progressive step in the right direction. But these fields are not generating power for residents of the Palm Springs, California, area. The natural energy these turbines create is actually being stored and routed out of the area while generating profits for a company based overseas. Meanwhile, local residents continue to endure regular rate increases and astronomical electric bills from the local power company.

Are you catching this? The same power companies that strive to paint a negative picture of those who choose to generate their own power with renewable energy (and reduce their dependence on these companies) are generating renewable energy, charging consumers for the use of it, and then doing all they can to discourage consumers from doing the same for themselves!

Enough said on that matter, but I sincerely hope that the next time you see or hear negative press on solar, you'll consider that perhaps these individuals may not be irresponsible, anti-establishment types but, rather, may simply be taking control back from a utility company in order to keep more of their hard-earned money … and just maybe you'll think about taking back control yourself.

Once upon a time, people followed their favorite shows by listening to them on the radio. Then television arrived in all of its grainy, black-and-white glory. Then colored sheets of cellophane were draped over the screen to see it in "color." Music was played from vinyl (I'm happy that's coming back). Phone calls progressed from switchboard operators, to rotary, to touch-tone, to cordless, and to cellular phones that looked like bricks. "Family wagon" at one time meant an actual wagon. And computers?

With all of the advances we've gone through to arrive at the conveniences we enjoy today, why should energy production and consumption be any different? The answer is: they're not. If we can learn to find our favorite show among hundreds of channels, download our favorite music from the Internet, use smartphones, drive hybrid cars, and store terabytes worth of data on thumb drives—why

> does it make sense to willingly subject ourselves
> to grid captivity? Over the course of generations,
> society has built muscle memory in cadence with
> technological advances. That technology already
> exists for solar energy, and America is lagging far
> behind the pace of advancement.

I'll spend the rest of this book showing you why and how to do that. If you'll join me for the short journey, I promise that at the very least we'll take the mystery out of solar, give you a few things to consider, and give you a "first look" at a future where you will be able to own a power grid for significantly less out-of-pocket expenses than what you would have paid at the meter ... and you'll be able to own it, leverage it, and save with it right where you live.

I'd put my money on the sun and solar energy.

—THOMAS EDISON

CHAPTER TWO

BUSTING MYTHS

AN ABBREVIATED HISTORY OF SOLAR

The first logical step in making a case for microgrids as the next solar renaissance is to take a look at how solar arrived at this threshold of energy independence. The journey from solar energy origins to modern technology is actually very interesting. Maybe I'm biased, but at the very least it makes for great material to impress your family and friends at your next dinner party.

THE EVOLUTION OF SOLAR

The legend of Archimedes' solar-powered laser dates back to the second century BC. According to the US Department of Energy, solar technology's history can be traced back to the seventh century BC, when the sun's rays were concentrated through a magnifying glass to make fire. By the sixth century AD, folks were using large, south-facing windows to create sunrooms in homes and public buildings.

In 1767, a Swiss scientist named Horace de Saussure built the world's first solar collector. The "photovoltaic effect" (PV is how electricity is produced using solar) was accidentally discovered by a French scientist named Edmond Becquerel in 1839.

In 1860, a French math professor named Augustine Mouchot began to develop the bright idea of turning heat from the sun into usable, mechanical energy. In his book *La Chaleur Solaire et ses Applications Industrielles*, he wrote:

> One must not believe, despite the silence of modern writings, that the idea of using heat for mechanical operations is recent. On the contrary, one must recognize that this idea is very ancient and in its slow development across the centuries it has given birth to various curious devices. (Mouchot, 1869)

Mouchot went on to present the first solar steam machine in 1878 at an exposition in Paris.

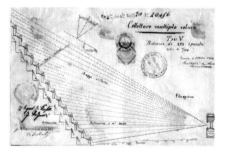

Alessandro Battaglia's patent application for a "Collettore multiple solare" (Multi Solar Collector), was registered in Genoa in 1886.

It was around this same time in Italy that my ancestor, Alessandro Battaglia, decided to improve on what Mouchot had started. His take on Mouchot's work was positive, but Alessandro saw inherent limitations in Mouchot's design. In response, he wrote and illustrated a paper titled "On the Methods and Convenience of Using Solar Heat for Steam Engines," which he presented at the Encouragement Institute of Naples in 1884. In it, Alessandro wrote:

It is not possible to build boilers of sufficient capacity to power industrial engines due to the fact that the boiler is mounted on the tracking collector, which limits its size; The boiler loses its heat easily because it is exposed to open air and cannot be insulated and protected; The tracking collector, as a single surface, is also limited in its total area. (A. Battaglia, 1884).

His solution (in simple terms) was to separate the boiler from the collectors, stretch it out horizontally, and then insulate it in a brick oven with a window that captured reflected light from over 1,200 flat solar collectors: one of the earliest examples of how energy could be managed with the addition of a storage device, in this case the

boiler. Realistically, it was a really early microgrid. Battaglia's patented "Multiple Solar Collector" system was said to have generated 50 hp (37.3 kW) and cost more than half a million dollars by today's standards. Pretty good output for the time, but finding a way to make solar affordable and practical for everyone was still a long way off.

In 1904, one of the greatest minds in history stepped into the solar fray. A young Albert Einstein published two theories that

Contrary to what is commonly believed, Einstein did not get his Nobel Prize for special relativity but for his contributions to theoretical physics including the photoelectric effect.

year: one was Einstein's general theory of relativity, and the second was what actually earned him his 1921 Nobel Prize for Physics—a paper on the photoelectric effect on which modern solar technology is based.

Demand for solar in the United States really began to take off around 1947 when demand for passive solar construction was high following World War II and the energy drain it placed on the nation. And solar became the preferred energy source in space in 1958 when the Vanguard, Explorer, and Sputnik were launched into space with PV wrapped on the outside to power electronics in lieu of heavy batteries. Many are still functioning decades later.

By the early 1970s, solar was starting to power homes. One of the earliest of these houses was "Solar One," constructed in 1973 by the University of Delaware. This structure had roof-mounted arrays, purchased utility power at night, and fed utilities surplus power collected during the day.

In 1982, the first solar-powered car drove across Australia from Sydney to Perth in 20 days, besting the time of the first petrol-powered car to make the same journey by a week and a half.

The use of solar energy to create steam originated with the Kramer Junction solar field in 1986. Five years later, in 1991, President George H.W. Bush ordered the establishment of the National Renewable Energy Laboratory, doing his part to demonstrate that support for solar and renewable energy is a bipartisan affair (as it should be). Another Republican, Governor Arnold Schwarzenegger, proposed an initiative for one million solar roofs by 2017, and President Barack Obama ordered that solar panels be installed at the White House in 2010, reflecting the dramatic rise in domestic solar PV systems through the 2000s—most especially in the United States, the United Kingdom, Germany, Japan, and China.

> *"Do as I say, not as I do."*
>
> *You've likely heard this phrase before. It's typically uttered by one person observing hypocrisy in another—most commonly by children in reference to their parents or subordinates in criticism of authority. Ironically, the statement will almost never be made by the one who is being scrutinized, because that would be an open acknowledgment of their hypocrisy.*
>
> *On one hand, utility-funded propaganda flies around the news and Internet, warning of the imminent danger that private solar users pose to the stability of the grid. On the other hand, utility executives recognize, and see as imminent, a smart grid powered by solar plus storage microgrids. Puzzling, right? You can read all about this by simply searching for* Disruptive Challenges: Financial Implications and Strategic Responses to a Changing Retail Electric Business.

In 2013, the world's largest solar-thermal plant, named "The Ivanpah Project," began generating renewable energy in the Mojave Desert thanks to 350,000 solar mirrors that cover approximately five square miles. The facility is made up of three separate plants that supply PG&E and Southern California Edison enough electricity to power 140,000 California homes during peak hours of the day. In addition to the energy generated at Ivanpah, the complex is said to avoid the equivalent of more than 400,000 tons of carbon dioxide (CO_2) emissions per year—a welcome reprieve for Californians, I assure you.

Arco was an early pioneer in large scale solar.

As of the writing of this book in 2015, the history of solar power has just hit its largest milestone to date with the completion of the "Topaz" in California's Carrizo Plain. The solar farm hosts nine million solar panels that cover over nine square miles and will generate 550 megawatts (MW), making it the largest solar farm on the planet. Like Ivanpah, Topaz will supply enough solar-generated electricity to power 160,000 California homes and will displace approximately 300,000 tons of CO_2 per year. Just around the corner, there are two additional solar farms expected to be operational in 2015: the 550 MW "Desert Sunlight" solar farm down the road in Riverside County and SunPower's dual-location project "Solar Star" that will generate 579 MW of solar-powered electricity and displace approximately 570,000 tons of CO_2 per year. To put the environmental impact in perspective, the emissions displaced by Solar Star will be the equivalent of removing two million cars from California highways over 20 years.

Ivanpah, Topaz, Desert Sunlight, and Solar Star are fantastic examples of the progress of solar and the viability of solar thermal and solar photovoltaic energy for the masses. In case you missed it before, they are also glaring examples of how power companies (PG&E and Southern California Edison) are relying on solar to generate electricity while resorting to scare tactics in the media to dissuade you from leveraging the power of the sun for yourself. While I join others in applauding these and future initiatives for their role in migrating away from nonrenewable, dirty energy sources, the fact remains that for all of the power that these solar farms are and will be generating for decades, those using the electricity will still be paying PG&E and Southern California Edison high retail rates for the use of that power.

And the electric utilities in California are rushing to complete their Renewable Portfolio Standard (RPS) with distributed generation—solar power on our rooftops. It's this kind of large-scale version that will make up the majority of solar's portion of California's new "50% by 2050" RPS policy.

In fairness to the power companies, there is substantial cost involved with the creation of these massive solar installations and the provision of the energy they supply. So if you rely on the middleman to provide you with energy for lights, hot water, cold beverages, and ice cream—and let's not forget your computer and smartphone—there's no room for you to cry foul. As I stated earlier, there's a reason the power companies don't want you to choose solar—and it has nothing to do with you putting the grid in jeopardy. It's bad for business, plain and simple. As if you needed further evidence, here are some proposals on the table from major California energy companies (PG&E, SDG&E, and SCE) that boldly illustrate their desire to kill the solar market.

The utilities submitted proposals in early August 2015 for NEM 2.0 when the current agreement ends in 2017. As CalSEIA put it, here are "lowlights" from the utility proposals:

» Each of them proposes to change NEM credits from full retail to a wholesale rate. PG&E proposes to set the compensation rate at 10 c/kWh, SCE at 9 cents, and SDG&E at 4 cents.

» PG&E and SDG&E propose residential demand charges of $3/kW and $9/kW, respectively. SCE proposes a monthly charge of $3 per installed nameplate kW of solar.

» PG&E proposes to change to monthly true-up, paying the net surplus compensation rate of 4 c/kWh for excess credits in any month rather than letting customers carry credits month to month. The justification: "This will simplify the program for our customers, which they have indicated they want."

» Each of them would limit virtual net metering and NEM aggregation.

As SEIA stated, "Never before has it been so clear, in black and white, that the utilities want to kill the solar market. And we all know they have big technical and legal teams to push their agenda."

Despite their efforts, the reason we've seen such a dramatic rise in solar over the last decade is directly related to the dramatic decrease in the cost of solar. For example, the Environmental Protection Agency (EPA) reported that from 2004 to 2011 the cost of solar dropped 72 percent from $4.00 per watt to $1.13 per watt. During that same period, the exponential rise in the cost of fuel increased 68 percent per mile for drivers, with fuel economy for light-duty vehicles only improving 18 percent over that time period. And because of the efforts of the utilities, the demand for microgrids and the subsequent cutting of the cord connecting us to them will only be accelerated.

Solar has come a long way in its history—but its usage (and efforts to learn and improve on existing technology) has always been about quality of life. The United States of America was founded for that very purpose. Nearly 240 years after our forefathers adopted the Declaration of Independence, there are over 300 million people living in the United States. But there are little more than a half-million homes relying on solar for power. That means the vast majority of Americans are choosing to be dependent on the power companies for their quality of daily life.

In 1973, one of the pioneers of solar, Karl Wolfgang Böer, lead the design and build of one of the first grid-tied solar systems at the University of Delaware, dubbed Solar One. What I find especially admirable about Mr. Böer is that, way before his time, he designed Solar One to prove that solar can and will provide all the power we need.

His team built Solar One over 40 years ago—with no efficiency overlooked in the design features—long before Energy Star stickers came to the forefront. And perhaps the coolest of all: he wanted to prove that solar could look good while doing it.

You can learn more about this esteemed pioneer and Solar One at Karlwboer.com. It's a quick read and one that I highly recommend.

Solar One was one of the first grid-tied solar systems ever constructed when it was built in 1973 by a team from the University of Delaware led by Karl Wolfgang Böer, one of the pioneers in solar who is still working in the field. Permission for use by Dr. Böer.

Solar has never been more affordable than it is now. With the pending solar renaissance featuring microgrids, real power will be available to Americans like never before. Solar started with the simple creation of fire—and today it's used to aid in the assurance that quality of life will go uninterrupted while providing a healthier planet.

No matter what moves you, be it saving money, saving the planet, or even cutting ties and taking back control of your choices from those exploiting their power (literally and figuratively), microgrids represent a viable, accessible, and relevant alternative to the status quo.

Stick with me, and I'll break it down for you.

SOLAR FACTOID

Did you know that the US Department of Defense's annual energy budget is $20 billion, making it the single largest consumer of energy in the world?

CHAPTER THREE

HOT BUTTONS

THE "WHAT" AND "WHY" OF SOLAR

With a bird's-eye view of the history of solar and having taken a brief look at the current events with solar today, we'd better make sure we're on the same page regarding the differences in solar technology and how they work—beyond the panels on rooftops you've no doubt seen in your travels. Like any good coach, if I'm going to encourage and equip you to "get in the game" of saving money and the planet while taking control of your personal energy production and consumption, there's no better place to start than with the fundamentals.

So far, we've referenced more than one type of renewable energy. It's important that you wrap your mind around the basics of how they work—not only so that you'll better understand the relevance of microgrids but also so that you can easily show off your renewable energy prowess at the aforementioned cocktail party or dinner engagement! There are far too many renewable energy and energy-saving technologies to cover them all in detail here, but at our corporate headquarters (literally named "The GreenZone"), we proudly educate, train, certify, and encourage guests from all walks of life on how to live green, save green, and go get green with fees paid for referrals of friends and family interested in solar and storage.

Two of the most widely known ways of capturing and producing renewable energy, apart from solar, are wind (such as the wind turbines we discussed previously) and hydro-generated (water) power production. The last two elements—wind and water—when paired with battery storage can also be applied to microgrid technology, but for the purposes of this book, we'll keep our focus on solar. One other technology you may have come across that actually *is* solar

A hallway at GreenZone lit by solar tubes.

"powered" (but not our focus here) is solar lighting for your home. No, not the glowing yard lights that quaintly line yards across America but devices that replace artificial light inside your home during the day. These tubular daylighting devices (or TDDs for short) allow you to light your home or office with natural sunlight and are amazingly effective at removing the darker spaces, while also reducing your daily energy consumption.

Let's take a look at the two most common forms of solar-powered energy technology: concentrating solar thermal (CST) and solar photovoltaic (PV). Before we look at the differences, I want to acknowledge the very positive commonality between these two methods of harnessing the inexhaustible sun to produce clean energy. CST and PV represent what I (and many other like-minded individuals) believe should be the biggest *duh* factor on the planet. Please give me a moment to elaborate.

In our wonderful, technology-driven civilization, power is a requirement. That's an indisputable fact. We rely on power for safety, hygiene, food generation and preservation, transportation, education, employment, and entertainment, not to mention medical care, communications, and commerce, just to skim the surface.

In general, there are only two ways to generate that power: either you burn nonrenewable fossil fuels to create it, or you harness renewable, natural energy. Either we depend on natural energy that renews daily, reduces existing and future pollution, and will never leave us hanging, *or* we can continue digging holes in the planet to tap fossil fuel sources that will eventually run out, that force us to make a mess of the scenery when we mine for them, and that will continue to pollute the planet while we burn them. Like I said, "Duh!"

In regard to providing inexhaustible energy through harnessing the sun, while making the world a prettier, cleaner, healthier, and cheaper place to live in, both thermal and photovoltaic solar are ready and willing to carry the power load for a very demanding human race. That said, they are the equivalent of fraternal twins. They may come from the same parent, harnessing the same energy simultaneously, but they look and function very differently.

Our focus will be on photovoltaic solar (just refer to it as "PV" and you'll sound like a solar pro). PV is what you're typically seeing when you look at solar panels on roofs or large ground rack systems. I'll break down how PV works, how it has led to the next solar renaissance, and why you should join and contribute to solar enlightenment. But first, let's give solar thermal its proper due.

> *No man is an island, but every form of microgrid can be. The microgrid is not a thing; it's an interconnected series of distributed energy resources (DER) working together to produce a system larger than any of the individual components could be on their own. "Islanding," as it is known within the industry, is the ability of the microgrid to operate in closed-loop fashion, completely independent of the grid—whether or not the microgrid is actually located on an island.*
>
> *Since we're on the topic of solar islands, "anti-islanding" is not an extremist utility group crusade. It refers to the safety standards that are in place to protect electrical workers during a system outage. If the grid goes down, anti-islanding protocol is there to make sure that your microgrid doesn't send power down the line when workers are trying to get the grid back up and running.*

SOLAR THERMAL

A very simple way to think of solar thermal is to remember that it is based on heat. If you lay something out in the sun all day, it gets hot. In most cases, prolonged heat without reprieve or intervention is less than ideal. In the case of solar thermal, it's a very good thing.

There are two types of solar thermal: passive and active. Passive systems have no moving parts and require no additional equipment. These systems are relatively small in scale and are frequently used in the residential setting to heat water for the home (hot water on demand) and also to extend the swimming season in cooler climate areas by using the sun (undaunted by lower outside temperatures) to heat pool water during the day. Active systems (moving parts) are also used on smaller-scale projects, but they are most profoundly represented in the massive thermal farms that have been erected around the country. However, their reliance on precious resources such as water, as well as labor, make them a less desirable option than PV.

In simple terms, these solar thermal power plants are a clean version of fossil fuel burning. Solar panels follow and capture the sun, the sun heats fluid in the panels to very high temperatures, the hot fluid produces steam, and then the steam is used to create electricity.

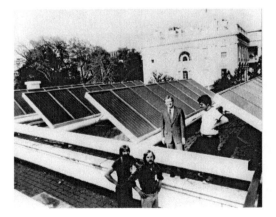

President Carter poses with the solar thermal system that was installed on the White House during his tenure there.

No pollution. No fossil fuels are required to generate the steam.

On this roof, a solar thermal system runs alongside a photovoltaic system.

PHOTOVOLTAIC SOLAR (PV)

The definition of photovoltaic is pretty straightforward: "Relating to the production of electric current at the junction of two substances exposed to light." You don't have to look far to find photovoltaic technology living in modern society. It's all around you—powering calculators, watches, parking lot lights, and thanks to the rise in electric-powered vehicles, electric vehicle charging stations. In the 1950s, photovoltaic technology was used almost exclusively for space—after all, there's nowhere outside of earth's atmosphere to plug in your garden-variety satellite.

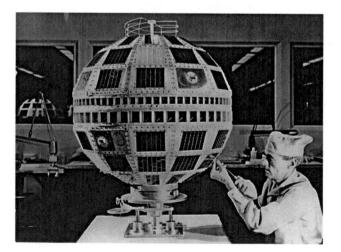

With PV having a long and successful history of reliability, practicality, and efficiency, it begs the rhetorical question, "Why isn't everything powered with PV?" In a word, the answer has been "cost." Power created by fossil fuels has historically been cheaper than solar technology, but as I've mentioned and will explain in greater detail in the next chapter, solar has never been more affordable and is becoming cheaper, yet more efficient, by the day. This broadening accessibility to PV coupled with the rise in costs of fossil fuel energy in cadence with the law of supply and demand has set the stage for the "dream" of a solar-powered society to become a reality sooner than later.

But how does it work?

There will be no quiz on this later—I have no intention of delving into the science of photovoltaics. Others much more learned than I am have written ample material on the energy that is found in light and the process of converting photons to electrons. What I will attempt to simplify here is how light from the sun ends up making your meter turn backward and, in the case of microgrids, will soon be providing you with your own miniature power grid, completely free from the utility meter.

Going back to the definition of photovoltaic, photo means "light" and voltaic means "electricity," so PV is simply the process of converting sunlight directly into electricity. How this process was discovered and developed through science and engineering is complicated, but mapping out how we get from A to Z is pretty straightforward.

Photovoltaic cells are made from semiconductor materials—silicon being the most effective historically and the most common. Cells are grouped together, linked electrically, and then presented nicely in a frame. This framed collection of modules, comprising cells that are made from semiconductor material, is what you're seeing when you look at a solar panel.

The cells absorb the sunlight, convert the energy to electrons, and then force the freed energy to travel in a certain direction, which is current. Using metal conductors on the top and bottom of the cell allows the current to serve as an external power supply.

Once you have current, you have to do something with it. If you want to be independent from the power company and its grid, you've historically had two options: use a backup generator for when your solar runs out, or invest in batteries to store the energy. Until now, batteries have been not only expensive but volatile, so most photovoltaic users have chosen a different option: tie your PV system into the grid.

By the way, if there's anything most folks have heard about solar, it's this "moneymaking" concept of selling electricity back to the power company. Is it possible? Yes, with restrictions. The policy of feeding the grid your extra electricity is called net metering. Net metering means energy produced by a privately owned solar system will roll the meter backward and provide a built-in "credit" toward electricity used outside of daylight and peak solar hours. At the end of the month, the utility customer is only charged for their net usage. According to the Solar Energy Industries Association (SEIA), " ... on average only 20–40 percent of a solar system's output ever goes into the grid—and that exported energy serves nearby customer loads" (SEIA 2015). This is your only option until you install your own battery storage solution.

Does it really work? Yes, people are doing it right now. There are many who see their meters roll back each month, free from their electric bills. But as I mentioned, it's becoming a hotly contested topic and being challenged by the utilities at this very moment. *And you are continuing to feed a grid and stay tethered to a conglomerate who pays you for your excess at the end of each 12 months at a wholesale rate but charges you for any excess at a high retail rate.*

Currently, the ability to go solar for no money down and the ability to save 25 percent or more on a typical electric bill—for life—is the reality of why most can, should, and do opt to lease their PV system. If you can afford to purchase a PV system large enough to cover your consumption and generate a surplus of electricity on a monthly basis, then a backward turning meter is certainly within an arm's reach. Add smart battery storage to your system, and you have yourself the makings of a microgrid!

> In your solar travels, you may hear the terms "in front of the meter" and "behind the meter" for describing solar systems. Unlike traditional systems connected directly to the grid (front of meter), behind-the-meter systems connect directly to the consumer's property, providing power that can either be used by the consumer or sent to the grid as generated, unused power. This is where net metering or meter "roll back" occurs. So when you hear the term "behind the meter" solar system, it just means the system is acting like an independent power plant, fueled by the sun that serves the consumer first—not the grid.

To wrap up this schematic description of a typical PV system and how it works, there are other components beyond the solar panels and battery. The charge controller ensures that your battery isn't overcharged or drained too much. The inverter is what allows you to use the direct current (DC) power generated by the solar panels by converting it into alternating current (AC), the kind that your appliances, computers, etc., run on.

Here's what a basic PV configuration looks like:

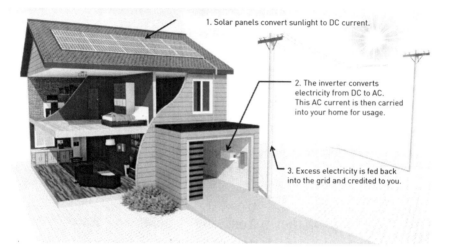

1. Solar panels convert sunlight to DC current.

2. The inverter converts electricity from DC to AC. This AC current is then carried into your home for usage.

3. Excess electricity is fed back into the grid and credited to you.

Sunlight hits cells ⇨ Cells absorb/convert energy ⇨ Current

Congratulations! You've just completed a very, very basic crash course in common solar technology usage and photovoltaic configuration—which leads us back to the main point of this book: microgrids.

MICROGRIDS

As I've referenced previously, a microgrid is a local grid that can be controlled and can operate connected to, or independently from, the grid. This is a very big deal for many reasons that we'll discuss for the remainder of this book.

First and foremost—at current capacities—the microgrid frees the owner (that's you) from the inherent flaw related to the interconnectedness of the grid. If you're 100 percent dependent on a grid for your power and the power goes down, you and everyone else connected to that grid lose power.

A microgrid means you have backup power when others don't. Microgrids are often tied to grids and will operate in sync with grid power levels until a power crisis arises—at which point the microgrid disengages from the grid and ensures you have uninterrupted power. Supplied with proper power by solar PV and battery storage, a microgrid has the capacity to run indefinitely and can be monitored and managed like many things we interact with and depend on every day—for example, a digital touch screen or mobile device.

Today's energy-delivery model consists of the utility companies charging consumers for energy used and tying them to an unexplainable monthly charge. But the model of being attached to a utility is not only dated, it's really one that we no longer need. From here on, we're going to take a closer look at why microgrids are such a big deal and how the current development of affordable and reliable battery systems will ultimately eliminate the need for grid dependence altogether.

I titled this chapter "Hot Buttons" because I know that understanding what solar is and how it works are not the motivators that will move you to go green, save green, and cut the cord entirely. No matter how impressive or useful something might seem, without relevance the end result is inevitably, "That's nice, but so what?"

So let's discuss what's in it for *you* when you choose to take control of your energy bill, hard-earned money, and a better future.

Save money. As I've already stated, the highest motivation for most folks is financial.

Solar saves you money. It's that simple.

If you lease your system, you can have a custom-built, attractive, state-of-the-art solar system installed on your home for as little as nothing down in many cases. Look for a solar company that can provide you with a fixed monthly payment with as high an offset percentage as possible to insulate you from ever-increasing utility rates. Credit requirements vary, but these lease options were designed to be accessible for most people. It's important to note that, with leasing, net metering is not an option, because you'll actually be paying the leasing company, not the utility company. Leasing also removes the need to apply for tax credits because the credits you would have received on the system are taken by the leasing company and applied to the cost of your system. If you have the discretionary income to invest in purchasing a system, then tax credits, net metering, and varying periods for return on your investment are all in play. If you purchase, your overall savings are the greatest. Either way, "go green" with solar and you'll start saving green from day one.

Save the planet. There have been a number of research polls conducted to identify what has motivated solar users to choose solar in the first place. I won't cite them here—a simple Google search will produce any number of articles on the topic. I bring this up because a very interesting fact surfaces—regardless of the source, the environment is a low motivating factor for many. Should that shock us? Probably not.

As a conservationist and one who loves nature and breathing clean air, I'd like to think that the environment is a high priority for people everywhere. In a volatile and recovering economy, however, it's clear why saving money ranks number one across all demographics. Still, the money savings are built into the solar system and happen passively. By choosing solar, you are also actively choosing to take control of your personal footprint and making a positive contribution to environmental cleanup and preservation.

SAVE BOTH!

Save brain cells. I'm a tireless worker. I demand nothing but the best effort, best service, and best results from those who work for me. One of the ways we work as a team to ensure that we deliver is that we strive to work smarter, not harder. I don't tolerate shortcuts or "mailing it in," but if there's a better way, a more productive way, a more efficient way that also happens to be a simpler and faster way, then I'm all in.

Which sounds more appealing to you: a watch that depends on batteries that wear out and must be replaced or a device that automatically charges with daylight? Would you rather drink from a mud puddle that will eventually evaporate, or would you rather take your refreshment from a clean mountain stream that constantly renews?

The battery is a marvelous invention, but what happens when you run out of batteries? If you're thirsty and there's no clean water nearby, a mud puddle could be a lifesaver, but it wouldn't be anyone's first choice. Fossil fuels are dirty, and they are running out—so much that wars continue to be fought over them. Sunlight is clean, never runs out, and requires no compromise or bloodshed to own. That's smarter, not harder.

Save budgets. Nothing wins more brownie points at the office than making or saving your company money. It's the reason why good salespeople and even better accountants and bookkeepers are in such high demand. What do most things in an office run on? After rent, payroll, and taxes, what do you suppose is one of the highest monthly bills for your company? What do you suppose would solve such a challenge—saving the company money and making you look like a star for recommending it? Enough said.

Save choice. Americans enjoy freedom in a way that most don't around the world. Freedom of speech, freedom of faith, freedom of choice—they're all part of the American dream. After all of my time spent on the front lines, it's still shocking to me how the vast majority willingly lays down its freedom to control their energy bills—whether it's from confusion, unawareness, or inexplicable indifference, my hope is that America wakes up and embraces its ability to save its right to choose control over misguided compliance.

Save lives. The same freedom I just touched on was, and continues to be, provided by the heroes who serve, sacrifice, and especially those who have given their lives defending and preserving freedom for every citizen.

I would never suggest that installing a solar system on a roof does anything to directly remove soldiers and all who serve their communities and country out of harm's way. But I can't help but wonder, as many have, *if there were no dependence or demand for fossil fuels, would there be a need for wars with those who would control it?* What we *can* know is that every human being who embraces alternative fuel and energy sources is one less person maintaining demand, and where there's no demand, supply becomes far less profitable.

An example closer to home is the danger faced by those living at risk from natural disasters, random "acts of God," and blackouts that leave people literally powerless when the grid goes down. Solar plus storage (in a microgrid) allows consumers to preserve life as they know it when the power grid fails them—and that is the difference between life and death more often than most people are aware of.

> Solar does more than contribute to better health through reduced emissions. In underdeveloped countries and rural areas where there is no access to power, solar can and is the difference between successful life support, surgery, and emergency medicine—or tragedy. The World Health Organization (WHO) published a 2014 study titled "Global

> *Health: Science & Practice," in which it reported that a quarter of all health facilities in sub-Saharan Africa are without electricity, and only six countries have facilities with diesel generators that either don't work or are too expensive to operate. Consequently, the WHO and nongovernment organizations and programs like the United Nations' Sustainable Energy for All initiative are turning to solar for a permanent solution that will literally save countless lives.*

In the following chapters, we'll talk about how microgrids will allow you to keep more of your money, help clean up the environment, and will play a role in fighting stereotypes—while redefining a generation. We'll also look at how microgrids can help make you a corporate rock star and how they will allow you to be part of *the solution* and leave a legacy for younger and future generations.

SOLAR FACTOID

The average taxpayer pays almost 100 times as much in subsidies for fossil fuels than for solar energy.

CASE STUDY 1

Ezra and Melinda Auerbach have had their own microgrid for more than 30 years on an island in British Columbia. Their original system was installed in 1982 with just batteries. Before that, in the 1960s and 1970s, Ezra was living "off grid" with kerosene and candles. He says, "It was nice not having a monthly energy bill then, and it's nice not having a monthly energy bill now."

Ezra, one of the founders of NABCEP, the North American Board of Certified Energy Practitioners, had his original system with just a couple of used Caterpillar starting batteries and a gas station-style battery charger with a generator that supplemented it. Lights were mostly propane with electric lights for "special occasions." Radio was good. TV was a little tricky.

Ezra and Melinda Auerbach have lived with their own microgrid for more than 30 years.

When they added solar, they also added their first inverter. It meant they didn't have to add fuel to the generator in the early morning to grind coffee. "When we got the inverter, it was magic," Ezra said.

And now? How has his system changed? The equipment and the personnel involved with installing the equipment are all more professional and code compliant. And now they have an SMA-based microgrid—three SMA inverters with 5.2 kilowatts in all and 750 amp hours at 48 volts, which gives them approximately 30 kWh of battery storage. His flooded lead-acid batteries are pushing eight years, and he's ready to move to lithium-ion, although the batteries prior to his current ones served him well for 13 years. The biggest change? How inexpensive solar panels have become. Their panels were $10/watt wholesale, which in the 1980s was a lot of money.

What is the most rewarding part of having a microgrid for Ezra and Melinda? "Watching the sun pour electricity into our batteries from the sky."

CHAPTER FOUR

KEEP YOUR MONEY

SAVING WITH SOLAR

Solar delivers savings the moment you flip the switch. If you lease, it's simpler than leasing a house or a car—you'll start saving 25 to 50 percent on your electric bill *immediately*. If you purchase a system, you can take advantage of the tax credits available to reduce the cost of your initial investment. You'll start saving on your electric bill immediately—once the system is paid off, you own it and the savings that come with it. Solar also *makes* money (beyond potential net metering) for homeowners purchasing solar systems because it adds to the home's resale value—current numbers being reported by the US Department of Energy are around $15,000 or $4/watt (Berkley Lab 2015).

A GROWING TREND

The first quarter of 2014 marked the first time in more than a decade that residential installations of photovoltaic (PV) solar exceeded commercial installations. It was also the first time ever that more than one-third of those residential installations came online *without* any state incentives. According to the Federal Energy Regulatory Commission Office of Energy Projects, in the first four months of 2015, 84 percent of new electric generating capacity in the US came from renewable sources. Of that, 25 percent was solar.

Those are impressive numbers that substantiate what I am sharing with you here. Homeowners are outpacing large companies with PV solar installations. One out of three homeowners is doing it without the help of their local government (who have incentive to support the utilities and corporations). Meanwhile, the utility companies are installing more PV solar than anyone, are subsequently charging you for the energy they are producing naturally, and—since I can't stress the point enough—are lobbying local government to take away your incentives while running scare-tactic public relations campaigns to dissuade you from choosing the same renewable energy they are leveraging to keep up with demand.

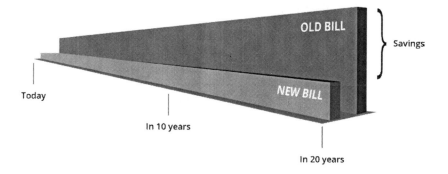

With a fixed rate on a solar lease, and continually increasing utility electric rates, savings grow exponentially each year.

All of that might get you fired up, or maybe it's interesting but not yet relevant for you. If you're ready to learn more, then there's no better time than now to make a call. After all, it costs you nothing to find out how much you stand to save each month.

That said, there is still much to share with you on how solar can help you keep more of your hard-earned money—and that's relevant to all of us.

Let's take a look back at why there is a common notion that solar is too expensive and why that is no longer the case.

PROGRESS OVER TIME

In a word, "technology" is at the heart of why PV solar used to be much more expensive than it is now. Technology always costs more to create in its earliest forms because, whatever that new technology is, no one has created anything quite like it before. It takes more work, and there's a process of trial and error required to optimize performance with the latest tools and materials available—and there typically isn't yet demand for it on a large scale. So technological pioneers (no matter how noble and forward-thinking their work may be) are working against existing limitations and simple economics.

It was true with computers, phones, entertainment, science, and the list goes on. So it was with solar. Part of that process was improving how efficiently PV solar panels were able to convert sunlight into usable energy. The fact is, while the sun may produce enough energy to satisfy our energy needs for all time, until the unlikely day comes that "the modern machine" runs exclusively on light, heat, and fire, we'll have to continue working on getting better at turning natural energy into readily usable energy. It should be noted that fossil fuels only deliver about 50 percent efficiency in delivering energy—this is not a solar-exclusive challenge.

In the early 1950s, PV solar panels converted less than 5 percent of the sunlight collected into usable energy, and the massive panels used to collect it were roughly 200 square feet in size. That's a lot of material to buy for producing panels that don't deliver a lot of energy return for the investment. As a result, the cost of PV solar in the early 1950s was a little shy of $1,800 per watt.

Fast-forward five decades, and as recently as 2012, PV solar panels had tripled their conversion capabilities between 15 and 20 percent and had been reduced in size to about 15 square feet. Less expensive panel production costs, improved technology, and energy conversion equal a drastically improved PV solar cost of around $1.30 per watt.

The panels tested and rolling out today are about the same size (about the size of a 42-inch flat-screen TV) and are able to convert just under 25 percent of the energy received from the sun—for an estimated PV solar cost starting at $0.70 per watt. This means that a 300-watt panel alone at $0.70 per watt costs $210. With the panel's estimated 25-year life, its output would be an estimated 11,000 kWh of energy at just $0.02 per kWh. Today, when you add in all the other equipment and installation costs in a turnkey residential solar system, your cost per kWh is less than $0.13/kWh. In 2012, that cost was $0.17/kWh. My point is, you shouldn't be looking at an overall system cost but rather your cost for each kilowatt hour produced by your system. Even if we pay no attention to the harmful effects of introducing carbon into the atmosphere, solar can, or soon will, compete toe-to-toe with the lowest market cost for fossil-fuel-generated electricity throughout the world. This means that rooftop solar-produced power is now less expensive and continuing to decrease in price more each year than utility-produced power in many parts of the world, including most southwestern states of the US. While the process of reaching cost parity everywhere else in the US is well underway, the fact is that burning dirty coal and exploiting other nonrenewable energy sources continues to be acceptable, rather than taking steps forward by converting renewable solar energy into usable power to satisfy the high demands of a consumption-driven culture. And when you take the environmental costs into consideration, solar wins every time.

Let's talk bottom line—solar is already less expensive than dirty energy that is produced and delivered by profit-driven utilities in California and many other southwestern states. There is no need for further discussion. For everyone else, does that mean that you should wait until solar "goes on sale?"

Not at all. And here are a few reasons why:

 » nonrenewable energy is dying

 » subsidization potential

 » common misperceptions

 » lost savings and income

 » Let's take a closer look at these points, and I'll explain.

IMMINENT DEATH OF COAL

Coal has long carried the lion's share of responsibility for energy production in the United States, but it's estimated that coal production will reach terminal decline in production within the next 20 years. This must be a reason why utilities are currently driving the rise in PV solar installations—if you're under the age of 60, they'll likely be running out of their primary source of electricity in your lifetime!

SUBSIDIZATION

If we put on our "optimist hat" and look at other countries that have embraced solar, we can see that *if* our country followed suit in subsidizing solar the way it does fossil fuels, solar would already be cheaper than the grid. Germany sees nearly 4,000 percent less sun than the United States but subsidizes solar and has installed 6,000 percent more solar than the United States. Currently, solar is cheaper than grid electricity in about 14 percent of the country, but if solar received the same subsidization money as fossil fuels do in this country, solar would be cheaper than the grid across the map. Wishful thinking? Examples like Germany show us that it is possible.

> *"Wait a hot minute," you might be saying. "Didn't I hear or read something from the Taxpayers Protection Alliance (TPA) talking about $39 billion in subsidies that the solar industry received?" A lot of media have picked it up and run with it, so it has to be true, right?*
>
> *A lot of good but misinformed people read or listen to talking heads reporting this garbage and have two*

choices, as they do with any news: accept it as fact or do their own research to learn if it's true or not.

Allow me to enlighten you on this report. First, no one actually knows how the TPA report came up with a $39 billion figure because it doesn't say. Second, of the 26 references cited in the report, 16 of them have either been founded by or received funding from the Koch brothers, who are well-known and highly publicized supporters of utility companies, state and US legislature lobbyists, and other prosperity-driven organizations—for example, Americans for Prosperity (AFP).

The Koch brothers are also behind the TPA. If you're doing the math, that means that the Koch brothers, who have interests in and financially support utilities, are the ones who paid for the report, paid the companies making claims against solar within the report, and paid for the TV smear campaign.

And speaking of the Kochs, here's my offer to them, as they happen to be our neighbor in the desert. Allow me to install a solar plus storage system so you can understand firsthand the benefits of a microgrid. I suggest that soon after your portfolio will expand to include solar, storage and microgrid technologies.

MISPERCEPTIONS

The majority of Americans think solar costs more than it does, but most would choose PV solar if cost was not a factor. That's a lot of people who are not using solar due to a misperception.

Many people also believe it takes longer than it does for solar to pay for itself and don't realize how much money they could actually be saving with solar right now—and how much more will be saved as costs continue to fall and technology advances.

If you compare solar to other dependency pain points—for example, fuel to run our vehicles—the savings are even more dramatic. Did you know that if you drive an average of 12,000 miles a year for the next 50 years, getting 20 mpg—while paying an average of $3.50 per gallon—that pencils out to $275,000 in lifetime fuel cost? In contrast, an electric car powered by PV over that same time period will cost you approximately $12,000.

LOST SAVINGS

There are very real, very significant, and highly relevant savings available to you right now simply by choosing to include solar in your daily life. These savings will only continue to grow, so choosing to wait means you're deliberately forfeiting savings while continuing to pay the utility company for electricity that is, or will be, generated by the same PV solar technology available to you directly.

LOST INCOME

As noted, there are a number of ways that purchasing a solar system actually earns you money in addition to the money it saves you. The act of purchasing solar comes with tax incentives including the federal Solar Investment Tax Credit, which is a 30 percent credit available on residential and commercial properties until the end of 2016. The aforementioned net metering is also very real and happening now for many who purchased or leased their solar PV system.

THE MICROGRID CONNECTION

As battery storage technology rapidly pushes the curve to become as efficient and affordable as solar panels have become, microgrids will allow you to generate and sustain your own energy—completely independent from the grid—and the amount of money you'll save on electric bills will increase with technological progress. Solar PV is a static investment. Battery storage now adds flexibility to that investment. Advanced energy systems are also future-proofed by nature in that they are modular and allow you to add to or upgrade components as more advanced batteries are developed.

The Galvin Project estimates that Americans pay *$150 billion* or more per year to cover the costs of power outages (Galvin-power.org). Microgrids provide stored solar electricity to keep you up and running when utility-supplied power goes down. Local power generation with PV and microgrids is also more efficient and can save you money by reducing the distance the energy has to travel, which reduces the energy loss that typically occurs because of transmission over longer distances. Microgrids can also reduce rates from the power company by helping to alleviate congestion during peak hours. If for example, you are on a time-of-use system, and electricity is more expensive between the hours of noon and 8 p.m., but your solar system's maximum production time ends at 6 p.m., you can simply discharge your batteries between 6 and

8 p.m. to subsidize or supply the power that is needed. This helps everyone, as it takes the strain off of an overburdened, antiquated grid delivery system.

> *If you happen to be one of our neighbors in Southern California and are paying Southern California Edison (SCE) or Imperial Irrigation District*  *(IID) for your electricity, and you haven't already calculated how much you could save with solar, I invite you to use Renova Solar's free solar calculator at www.renova. solar/calculator. php.*
>
> *It could be a real eye-opener for you.*
>
> *Elsewhere in the country, www.pvwatts.nrel.gov can be a useful tool to estimate the energy production and cost of energy for grid-connected PV energy systems—as a first step to energy independence.*

I am passionate about serving others—especially when my service allows you to keep more of your money so that you have it to invest in whatever *you* happen to be passionate about.

Most folks I talk with—who haven't taken a serious look at solar—have hesitated because they are unclear regarding how affordable and accessible it has become, and they have no real idea of how much there is to lose by relying on a defective, overpriced, and outdated power grid system. I've made the best case I can for the monetary savings, and I'll leave it to you to exercise your freedom to investigate and choose what is best for you.

In addition to helping others pursue their passions, I am also passionate about cleaning up the mess we've made of the planet so that our children and future generations won't be forced to live in the sort of world we're heading for. I don't care what side of a political line you live on. I've never met anyone who thinks a brown sky is better than a blue one or that an industrial plant is prettier than a green, growing plant with flowers on it.

Let's dive into the positive environmental impact of solar … no tree hugging required.

SOLAR FACTOID

Solar energy is the main source of energy for all life forms.

CHAPTER FIVE

HELP CLEAN UP

THE ENVIRONMENTAL COMPONENT

Not everyone is passionate about the environment. I get it. It's often a polarizing topic that has drawn more lines between us than it has built bridges. My motive for addressing the environmental benefits of solar energy here are not politically motivated, as I believe the efforts to address pollution and its effects on climate change affect us all, and my sincere desire is that policies like the newly proposed Clean Power Plan don't get caught up in partisan politics. As a pragmatic member of the Republican party with very socially liberal beliefs, I distance myself from those in the party who deny that climate change is real and are preparing to lead a fight against efforts to combat it. The Clean Power Plan, announced by the Obama administration in August as this book heads to print, has the EPA giving each state an individual goal for cutting power plant emissions that must be met by 2030. At its completion, the EPA expects emissions that are 32 percent lower than they were in 2005. As the folks at Vote Solar (VoteSolar.org) said, "For too long, emissions from polluting power plants have been a quiet scourge across the country, causing illness and taking lives—especially among our nation's most vulnerable."

The emissions limits will also boost prices for traditional power such as coal, oil, and "nat gas"; make solar even more attractive; and cause states to offer additional subsidies to help encourage rooftop solar in an effort to reduce demand from fossil-fuel-burning electric power plants. Of course, all of this can change based on who next occupies the White House, how the courts rule on any challenges, and how the states react. I'm hoping this plan galvanizes your efforts to support big ideas like this, starting with your own energy independence.

It's a big step forward, but I'm not here to pander to you as if saving natural habitats, resources, and the planet will be the hot button that compels you to take action. I hope that those things matter to you, but as we've found, it may only be a secondary motivation for you—and that's okay.

That said, nearly every person I've met between the ages of 25 and 51—individuals labeled as members of Generation X and the first millennials—have been more aware of and consistently hold more appreciation for the breathtaking beauty that can be found

around our planet. Each year, the millions of people who flock to American treasures—the Grand Canyon, Yosemite, Yellowstone, Glacier, Niagara Falls, the Hawaiian and Virgin Islands, and the fjords and Northern Lights of Alaska—can't be completely apathetic to the cause of wanting to keep these beautiful places beautiful.

Whether your support for the preservation of such destinations is limited to the money you might spend to visit them or if you regularly support and participate in such preservation efforts, choosing to claim your right to freedom and your own money are still likely the best reasons to go solar. The fact that you can actually make a very real difference is an added benefit whether that fact drives you or not. Let's look at a few ways that you'll be "going green" while saving (or even earning) more "green."

EFFORTLESS CLEANUP

The average American household produces over 16,000 pounds of CO_2 emissions each year, and the number-one source of these emissions is energy. When you install a 6kW residential PV solar system, the reduction in your personal carbon footprint—over 25 years—is the equivalent of not driving (or flying) hundreds of thousands of miles, actively planting thousands of trees, and eliminating the trash and waste of hundreds of people.

Stop and think about the effort it takes to drive for more than 400 hours. Now think about how much patience it takes to sit on an airplane for 30 hours. Can you imagine planting over 300 trees by hand? Does the thought of removing the trash for a dozen people every day for an entire year appeal to you at all? Do all of that *every year for the next 25 years,* and you'll accomplish the same as what your PV solar system will be accomplishing in reduced emissions simply because you chose to install it and save yourself 25 percent or more for life on your electric bill. That's green times two!

> *Storage batteries (like solar panels and other components before them) have taken great strides toward greater efficiency, longer performance, and lower costs—for instance, the batteries we have developed and continue to improve through our MYCROGRID project will ultimately allow for older storage battery units to be recycled or repurposed and replaced with higher-density models.*

ELIMINATE THE PUMP

I've already touched on the financial savings you'll enjoy by driving an electric vehicle powered by PV solar. The list of available electric vehicles is growing each year. Now that EVs and hybrids are no longer a niche or fad, it's likely you won't have to leave your favorite brand to own or lease one, and the competition is driving down the up-front cost for most makes and models. Coupled with a personal PV system, the growing number of electric charging stations that power up your vehicle while you shop means that you can quite literally *stop paying for gas forever.*

The environmental impact and health risks associated with fueling stations have been known for years, compelling lawmakers and government agencies to increase regulations for improved safety. Even without statistics, common sense dictates that spilling oil and gas on the ground can't be good for the soil, the air, or the people working at and living around these 24-hour oil-vending centers. So skipping the pump is not only good for your wallet and the air quality, but it also serves to eliminate the need for (let's face it) what are often the dirtiest and most uninviting corners in society.

Speaking of oil spills—I'd like to go back in time to touch on a spill of historic proportions.

The year was 2010. British Petroleum (BP) had a doozy in the Gulf of Mexico. In fact, it was the largest oil spill in United States history, with 287 million gallons of crude oil pumped into the Gulf over the course of nearly three months—affecting over 16,000 miles of Texas, Louisiana, Mississippi, Alabama, and Florida coastline. Oil is still washing up on shores to this day and poses health risks to residents. Even more tragically, the oilrig explosion that caused the spill claimed the lives of 11 people and injured 17 others.

BP has reportedly spent over $14 billion as of 2013 to clean up the oil spill. BP also spent over $500 million in a massive public relations initiative for brand repair. Following a 2015 court ruling, BP will be responsible for a reported total pre-tax charge of $53.8 billion including up to $18.7 billion in penalties, and the company has sold off roughly one-fifth of its assets to pay for the tragedy (Reuters, July, 2015). Besides the obvious point that bad things can and have happened as a result of America's dependence on nonrenewable energy, there's a solar connection beyond the fact that dependence on renewable energy would ultimately eliminate the risk of disasters.

To put the spill in perspective, it was the size of the entire state of Kansas—and it cost BP over $30,000,000,000 to clean it up. If you covered the state of Kansas with solar panels you'd produce enough natural energy to power the United States—and Central and South America—for the next 25 years. The amount of oil that BP spilled only offered enough energy to power the same area for *one* day!

A typical 5kW solar system:

Eliminates the equivalent of more than six tons of CO_2 emissions annually.

Cuts the equivalent of 25 gallons a week in gas consumption.

Equals the oxygen regeneration of planting more than 2.5 acres of trees.

THE MICROGRID CONNECTION

On the most basic level, I call microgrids the next solar renaissance because they are the next logical step in PV solar, renewable-energy progression. While solar currently only generates around 1 percent of energy in the United States, it was recently reported by the US Energy Information Administration (EIA) that PV installations have grown over 400 percent since 2010, and all signs are pointing to a rapid migration to solar. Recent tragedies and current events are providing highly compelling motives to take a hard and immediate look at solar.

In 2012, Hurricane Sandy, "… destroyed whole communities in coastal New York and New Jersey, left tens of thousands homeless, crippled mass transit, triggered paralyzing gas shortages, inflicted billions of dollars in infrastructure damage and cut power to more than eight million homes, some of which remained dark for weeks" (*New York Times* 2012). Demand charges for companies operating on the grid are on the rise. Clean air regulations are forcing the closure of numerous coal plants. It's no wonder that public utilities are facing a growing vote of no confidence from residents and consumers alike.

University of California, Riverside has recently launched Sustainable Integrated Grid Initiative (SIGI), one of the largest renewable energy project of its kind in California. This initiative has been developed specifically to research the integration of intermittent renewable energy, energy storage, and all types of electric and hybrid electric vehicles.

The Sustainable Integrated Grid Initiative key features include:

- *4 megawatts of solar photovoltaic panels*

- *2 megawatt-hours of battery energy storage*
- *27 electric vehicle charging stations*
- *An electric-powered trolley service*
- *Energy monitoring and smart dispatch*
- *Open architecture designed for expansion*

One of the challenges of PV solar historically has been the loss of independence once the sun goes down. Enter solar battery storage and the microgrid. The catch is that battery storage has remained cost-prohibitive despite the steady decrease in cost and increased accessibility to the PV systems generating the power they are storing. This challenge is being faced head-on, and great strides are already in the works.

I'll touch on the advances in battery technology that we're a part of with our Mycrogrid project before this book is over, but I've got my eyes on research and development taking place on both coasts. At the Pacific Northwest National Laboratory (PNNL), research and development is underway on a sodium-beta liquid battery, and at MIT, Group Sadoway's liquid metal battery (LMB) project is developing a low-cost battery with a long life span. Both projects are targeting grid-scale stationary energy storage. At the Bourns College of Engineering Center for Environmental Research and Technology at University of California Riverside, the Sustainable Integrated Grid Initiative is researching the integration of intermittent renewable energy, such as photovoltaic solar panels; energy storage, such as batteries; and all types of electric and hybrid electric vehicles. It's the largest renewable energy project of its kind in California. This is in addition to ongoing work to develop high-performance solar cells and enhanced battery and thermal storage technologies.

Just as batteries have advanced from powering small electronic devices to now empowering vehicles to drive for long distances on a single charge, the evolution of battery storage is a certainty because it is a necessity.

These advances in battery technology will make battery storage scalable to massive proportions and accessible to all—ending any debate over PV's viability as a total replacement for nonrenewable energy. It's going to happen, and it is people like you who will play a role in changing the face of the world we live in while establishing a legacy that will redefine a generation.

SOLAR FACTOID

Ironically, Saudi Arabia (yes, the oil exporter) has committed $109 billion to achieve 100 percent renewable energy in its country—think they may know something about their oil supply that we don't?

CHAPTER SIX

FIGHT THE STEREOTYPE

HOW ENERGY RELIANCE CAN HELP
DEFINE A GENERATION

As a solar expert, educator, and professional in California for more than a decade, I have to admit there have been times that the battle to raise awareness and inspire action has been perplexingly daunting. In a region with more year-round sunshine than any other in our country and one with notoriously high energy bills and power shortages, my fellow solar champions and I have been forced to ask the obvious question on a daily basis: Why aren't more residents taking advantage of the clear benefits of solar? Accessibility and cost are factors, but if you'll indulge me for a moment, I believe there has been a generational cause in play as well.

If you Google "generational differences between boomers, Generation X, and millennials," you will find a multitude of sources that tout various group comparisons—but for the sake of this discussion, let's agree that "boomers" are individuals who were born from 1946 to 1964, "Generation X" adults were born between 1965 and 1984, and members of the millennial genera-tion were born between 1982 and 2004, so there is some overlap and then some Generation Y in there that pre-dates millennials by a bit and runs alongside them.

Let's focus on millennials for a moment and then run through each group. Millennials are now starting their own households and making their own payments. And according to a 2015 J.D. Power survey of customers of 145 utilities that received 27,000 responses, millennials like to take control of their energy usage and are more likely to use software. Many agree that millennials are the main driver of the community solar movement, and as even more become homeowners, they will drive new concepts, such as smart homes and other programs.

Historically, the small fraction of the population taking advantage of solar has consisted of retirees and affluent professionals with the discretionary income (and fiscal awareness) to take advantage of tax rebates and who've known that the investment would pay for itself while they were still actively enjoying retirement. This group of savvy solar users has largely fallen into the "boomer" category.

Boomers are the generation that was promised the American dream and actively pursued it while revolutionizing the American way of life. Had solar been as effective and accessible in the 1960s as it is today, we might not be having this discussion. After all, you'd think that a generation known for its questioning of authority, distrust of "the establishment," and prevailing belief that "anything is possible" would have embraced a revolutionary alternative like solar.

Unfortunately, solar technology wasn't as effective or accessible then as it is now, and these cultural pioneers matured into professionals with a different awareness and perception of solar energy as a whole. As a result, there has been a need for a reeducation of sorts with boomers. We're making headway, but it's been slow.

Enter Generation X, consisting of "latchkey kids" who grew up largely having to take care of themselves during the rise of increasingly visual political corruption and who watched their parents largely fail to achieve the American dream. This generation is independent, self-reliant, understandably skeptical, and suspicious of their parents' values and ideals. That's a tough nut to crack when a universally positive alternative like solar has often been lumped in with other liberal/environmental causes that are rejected by about half of this generation. For solar advocates like me, this makes it all the more frustrating at times—knowing that solar delivers on the Generation X prerequisites of independence, self-sufficiency, and verifiable results.

Fortunately, members of Generation X are also entrepreneurial, globally minded, and technically savvy by nature. They are the ones who have largely brought solar to its current state of effectiveness and accessibility. They are also the ones who are leading solar into the inevitable future of microgrids, battery storage advancement, and global self-sufficiency in energy consumption. Unfortunately, the nature of prevailing generational paradigms—coupled with the fact that (more than boomers and millennials) Generation X has felt the blows of economic recession, bank collapses, and the housing bubble bursting—means that even members of Generation X, who know that solar is the right way to go and more accessible than ever, still aren't capable of or ready to take advantage of it. This generation is coming around now that home ownership and discretionary savings are no longer prerequisites to leveraging solar, but we're still fighting resistance caused by the lingering effects of circumstances that have only reinforced a preexisting tendency toward skepticism and caution.

If you're a boomer or a member of Generation X, I would say to you that even though it hasn't always been this way, solar is and will continue to evolve into a solution that speaks to your sensibilities, delivers on generational motivations, and makes more sense than ever—especially with microgrid technology and how it allows you to "baby step" your way into energy independence.

If you're one of the millions who are part of the generation known as millennials, then you are in the perfect position to play a role in leading a generation to cut the cord—just like the boomer generation was poised to bring about change in the 1960s—but with the benefits of advanced technology at your fingertips.

I'm not a millennial, and I don't pretend to understand everything about this generation. But my observations from working with members of Generation Y are that:

1. Dependence on technology is not a flaw—that's its purpose.

2. The propensity to work smarter and not harder is often misperceived as a lack of work ethic by those working much longer hours and struggling to produce the same results as their younger counterparts.

3. They question, challenge, and push back because efficiency is not only important to them—they demand it.

If any generation is going to take the solar plus storage ball and run with it, it's going to be millennials. I hope the rest of us "older dogs" can pick up a few tricks from them along the way. And if we do it right, we'll arrive with a sense of fulfillment and accomplishment—with no regrets.

As of 2014, millennials represent the largest generation in the United States—one-third of the population—and consequently, the largest consumer base with more overall purchasing power than boomers or Generation X. Millennials have been raised with technology while witnessing the injustices of the world from behind the shields of their disenchanted and protective parents. Millennials are the most educated generation in our nation's history and are keenly aware of (and responsive to) the stereotypes often attached to them by boomers and older members of Generation X.

Millennials are a generation that no longer lives with separation between life and work but rather pursues their passions so that work and life are increasingly seamless. It is a generation determined to turn around the "wrongs" that are seen in the world today. It is a generation that is actively establishing loyalty to and advocacy of brands, causes, processes, and solutions. For millennials, technological advancement is not suspect but expected. Optimism, openness to new ideas, and an even greater sense of global responsibility prevails among members of Generation Y. For these reasons and more, I believe that (more than any generation before it) the millennial generation is uniquely equipped, inclined, and ready for a convergence with solar energy and microgrid technology.

In fact, there may be no other cause/solution that shares the plight of millennials more than solar and renewable energy. Often prejudged, dismissed, or disregarded completely by older members of society—like solar—millennials are often forced to confront and overcome stereotypes and erroneous paradigms about their value and place in society. The good news is that solar and microgrid technology can actually play a role in redefining the millennial generation.

The millennial generation's dedication to challenging and questioning the status quo, and its drive for greater efficiency and productivity, are why Generation Y is playing an increasingly active role in the advancement of solar technologies and most especially the advancement of battery storage capacity that make the microgrid the next solar renaissance.

If you think about it, what better way is there to disprove a misconception of generational unproductivity than to be the generation that brings energy independence to an almost entirely dependent and utility-exploited society?

More than any generation before it, Generation Y experiences, understands, and demands the immediate and long-term benefit that advances in technology provide, which is why they are the perfect match for being the champions of solar and utility grid cord cutting.

Millennials are environmentally minded but largely refuse to be associated with the often negative label of "environmentalist." *Today's solar is not the same as it was, and it refuses to be defined by limitations of the past as it works to make a positive global impact on the environment.*

Millennials are eager to save money on things they need to buy so that there's more money to spend on what they *want* to buy. *Solar saves (and makes) money so that you have more to spend on what's most important to you. And when you add storage, you achieve personal, long-term freedom and maximum savings.*

Millennials are also more team-oriented and collaborative than their predecessors, which is why millennials will be the generation to drive the solar renaissance and deliver a brighter and energy-independent tomorrow—a tomorrow that they will benefit living in and one they can be proud of handing over to the leadership of future generations.

If you're a member of Generation Y (or know someone who is), I would also point out that embracing solar energy and microgrid technology will have a profound effect in the workplace and will help establish millennials as a generation of positive change and progress. While microgrid technology will continue to advance in step with battery storage capabilities, leveraging the power and efficiency that solar provides *now* will enhance corporate brands in a millennial-driven marketplace as "green energy" companies and can make those who facilitate the integration of solar power at the workplace look like corporate rock stars to boomer and Generation X senior executives.

A great example is the spontaneous formation of Black Rock Solar, a nonprofit solar design and installation organization that has its roots at Burning Man. Black Rock Solar got started in the summer of 2007 when a group of volunteers led by Matt Mynttinen and Bill Brooks installed a 30-kilowatt array on the playa surface in Black Rock City during the annual Burning Man event, themed that year as The Green Man. Later that fall, the team moved the array and installed it in the tiny desert town of Gerlach, Nevada, donating the solar project to the town's school system. The original array was expanded to 90 kW, providing the school with 30 percent of its current power needs and saving educators more than $15,000 per year. It was so well received and satisfying that an entire organization grew up around it, and Black Rock Solar has now installed 83 clean energy projects totaling more than 5 MW. Arrays have been built at schools, senior centers, hospitals, food banks, homeless shelters, and Native American tribes at little or no cost to the recipients. The installations currently provide more than $725,000 of combined savings annually for the organizations.

SOLAR FACTOID

As of 2014 it is estimated that the global solar electricity market is worth over $1 trillion annually.

CASE STUDY 2

Dubbed "Eagle One" because of their address and because it was the first residential battery storage system to be installed by Mycrogrid, the components in this case study include a 16 kWh SonnenBatterie Eco battery storage system that works simultaneously with an existing 8 kW DC Solar Photovoltaic (PV) SunPower System—previously installed by Renova Solar. This couple, retired and wanting to fix their expenses, expressed interest in having a system that would handle critical loads in the event of outages or a natural occurrence, such as an earthquake. Since this is a desert home, going without electricity for a short period of time could be extremely uncomfortable and possibly deadly. Mycrogrid engineered this particular system to connect the air conditioner, refrigerator, and Internet router even at the most critical loads.

If and when the grid goes down, the battery system will initiate and enable the homeowners to utilize the loads on this load panel for an extended period of time—up to ten hours without power from Southern California Edison's electric grid. Paired with the solar PV system, they will be able to continue to power their loads from the batteries and have the batteries recharge from their existing solar system. The advantage of the solar PV system is to have the ability to charge the batteries when the grid is down. On an ongoing basis, they have the ability to discharge the stored energy at any given moment and for any duration of time. This could prove to be very effective in

a time-of-use rate schedule where homeowners can choose to use the battery and solar power to energize their home when the price per kWh is most expensive.

The battery system is very quiet, radiates no additional heat from its structure, and is located in an interior closet in the customer's home—the batteries will remain cool and work at their utmost efficiency. There is additional space to install more batteries, if the homeowner is so inclined. The remainder of the electrical components, including the load panel, are located in the garage. As with all Mycrogrid systems, this one is future-proofed—the components are modular and can be updated if new and better technology is available. As markets for energy collaboration bloom, this family will be able to sell its stored power into the grid, making this a flexible, secure, dependable, and profitable investment.

CHAPTER SEVEN

CORPORATE ROCK STAR

SOLAR'S ROLE IN THE CORPORATE WORLD

Does it seem contradictory to advocate for separation from dependence on the energy establishment while promoting the advancement of solar energy in the corporate workplace? With historical references to solar titling it "radical," "a fad," or "cost prohibitive," it may seem odd to promote individuality and self-governance on one hand while inviting corporate America to join the revolution. But the truth is that solar dependence is rising to the top of the corporate food chain, and there is a growing body of evidence with case study data to support solar advocacy where you work—whether you do your thing in a corporate high-rise or operate from a small business venue. Allow me to illuminate.

As recently as 2014, it was widely reported that America's best and brightest companies were not only embracing but expanding their reliance on solar power. Google, Apple, Wal-Mart, Kohl's, Costco, and IKEA were just a few of the corporate giants playing their role to usher in the next solar renaissance. In late 2014, Verizon announced it was adding a $40 million investment to the $140 million it had already invested in the previous two years.

Verizon made no secret or apologies for its motivation to invest $40 million in solar: driving shareholder value. James

99

Gowen, Verizon's chief sustainability officer, made that clear in a 2014 Bloomberg TV interview when he was quoted as saying, " ... solar delivers value for us all year long ... Green energy—regardless of your industry—is good for all" (Lacey 2014). Even among residential solar users, benefits to the pocketbook (not an irresistible love and concern for the environment) are spurring the growing reliance on solar, and that's okay.

> I'm an avid practitioner of yoga for individual reasons, but whether someone does yoga to live longer, to look good in a swimsuit, to be a part of the social experience, for reasons of faith, or all of the above, the benefits of yoga are indisputable. The yoga mat doesn't care why you're on it, but you'll be better off for using it regardless of your motivation, and society as a whole benefits from healthier individuals.

I think most agree that if you can do something positive for the environment and make the space around you a nicer place to live in, then that's always a good thing. So whatever your individual reason to want to have clean, renewable electricity, that production benefits society as a whole as well. The Fortune 100 companies are also making an investment in solar. We should all rejoice because sometimes the technological innovations they employ ultimately lead the way for other advances that benefit every business. And what benefits every business can potentially be great for you—the individual working in a corporation—as the one who plants and waters the seed that will ultimately bring in a bumper crop for your business or company.

And let's be clear: The investments already being made in solar, and being a sustainable corporation, are in existing solar technology—the same technology available to you and your family, friends, and neighbors. So you're sure to win brownie points in your private circles—outside the office—for being a solar evangelist as well.

It's also important to note that awareness and discussion of microgrids is only just beginning now that initial technology has made it relevant and applicable for residential and corporate use. If you do decide to be the company's solar champion, then you'll need to know how to speak intelligently on the topic and on the next great advance in solar—an advance that even some solar providers are not entirely aware of yet.

I'm not one of those providers who is still in the dark; I'm a leading advocate for the newest advances in solar. I'm sharing my insights with you because I'm one of the relatively few in the solar industry who is currently working on the front lines of microgrid development, education, and integration. I'm proud of the fact that my company has been at the tip of the solar spear since its inception— as evidenced by our awards, national recognition, and successful initiatives that have resulted in great advances in the solar space.

You may have an opportunity to be the tip of the spear for your company as well. If your company owns its space and hasn't yet moved into the new energy economy, you can demonstrate your tech savvy, lead the charge in mitigating costs and protecting company assets, and show concern for others and environmental responsibility—all of which show what a visionary, proactive, and invaluable asset you are to your company. If your company isn't already thinking of making the shift to solar, (or still believes it can't) it's likely going to be a tough sell at first—take my word for it. But I'm providing you with as much information as I can to help you plant the seeds for renewable energy and a sustainable corporate environment.

To properly champion solar in the workplace, you'll need to know the well-established facts about the corporate benefits of solar, so I'll save you the initial research and lay them out for you here. The benefits of solar in the workplace include (but are not limited to):

» reduced operating costs

» positive return on investment (ROI)

» tax advantages

» reliability

» protection against rate increases

» improved net income

» improved reputation and public relations

» generational impact

The extent of these benefits will vary depending on various factors, but they are applicable for small and large businesses alike.

Let's look at each of these individually.

Among many commercial solar installations is one for Goldenvoice at the site of the popular Coachella and Stagecoach music festivals. Shown is Alan Roby with author Vincent Battaglia.

REDUCED OPERATING COSTS

Solar will reduce and can even eliminate a building's electric bill. It's been reported that for some this can equate to prepaying for almost 40 years of energy. That's a big deal for obvious reasons. Depending on where you operate, the cost per unit for the power your company is using is likely higher than what it can currently cost to install a solar power array for energy independence. If you eliminate your electric bill, you reduce your monthly cost to operate your business—pretty straightforward.

POSITIVE RETURN ON INVESTMENT (ROI)

When it comes to capital investments by your company, there are few that can match the positive return on investment (ROI) of solar. Net present value (NPV), internal rate of return (IRR), and the time it takes for payback on investment all make solar a wise capital investment in a time when caution and demand for maximum return are key for business decision makers. With decreasing cost and increasing efficiency, investing in solar delivers a relatively quick and solid long-term return.

TAX ADVANTAGES

The current Solar Investment Tax Credit (ITC) was extended in 2008 through 2016 and offers a 30 percent tax credit for individuals or businesses that purchase qualifying solar technologies. If you (or your company) have not looked into solar, you owe it to yourself to explore your options and get a quote from a qualified and reputable solar installation contractor. There is no guarantee that this 30 percent federal tax credit will continue beyond 2016.

The ITC and other available rebates and incentives effectively reduce the cost of your solar system by 30 percent or more.

RELIABILITY

Once microgrids are fully integrated into solar systems, battery storage of the energy produced by solar will no longer be an ongoing maintenance concern. But systems that operate without battery storage backup (the most common type of system) operate with no noise, mess, or fuss for 25 to 40 years. And panels from certified, reputable manufacturers are backed by warranties.

How many other pieces of equipment in your office can you say will work for decades with little or no need for maintenance or replacement?

PROTECTION AGAINST RATE INCREASES

Energy rate hikes from the power company are as sure as death and taxes. I've already pointed out how solar has the potential to help us all live longer, and we've just covered the fact that solar can currently knock 30 percent off of your tax liability. In the current utility regulatory environment, solar alone defends against those inevitable cost increases to keep the lights on, computers and office equipment running, food and drinks cold in the refrigerator, etc. while tied to the grid.

So will the competition for customers who are considering solar keep the power company from raising its rates? No. It's simple. Utility electric rates will continue to increase, especially investor-owned utility (IOU) territories, and we're already seeing power companies use loss of customers because of solar as justification. Solar alone will only protect your business from the utility rate portion of your bill and not future increased demand charges. However, with the addition of battery storage, your business is fully insulated. Depending less on your local electric company (or not at all) means your electric bill is less or nonexistent. That's how solar protects you and your business against rate hikes now and, with storage added, in the future as well.

IMPROVED NET INCOME

Under current conditions, as of the writing of this book, the best-case scenario is for your business to actually sell unused electricity back to the grid, which turns your company's investment into a money-making asset, known as a feed-in tariff. If you find that option is not available in your area or your company isn't capable of installing a system large enough to generate excess power to sell back, you'll still see an increase in profitability due to the reductions in operating expenses that solar delivers. An increase in profitability increases the value of your business, which makes your company more attractive for potential growth lending—and growth is what every company is looking for to some degree. Your advocacy for solar could be the catalyst for increased company value and growth. Adding battery storage allows business to get even stronger as demand charges are also eliminated because now, in addition to the power generated by solar, you have the flexibility of storing the lowest cost electricity (solar or the grid) for use to offset times when rates are higher. We refer to this as an energy arbitrage asset.

IMPROVED REPUTATION AND PUBLIC RELATIONS

The marketplace today has evolved even more rapidly than solar. Today, brand is no longer what you tell the consumer it is; your brand is what consumers tell each other it is. Consumers control an increasingly segmented marketplace where digital presence now rules, and online reputation is key to building your corporate tribe of brand advocates. Whether you are personally tied to the brand marketing and public relations for your company or not, ensuring that these areas are well managed and communicated mean the difference between being seen or invisible, being heard or tuned out, and being awesome enough to have your brand message go viral or languishing in the noise that is a saturated digital space.

Go solar and your company will reduce its carbon footprint, legitimately promote itself as a company that conserves resources for the benefit of the community, and reap the reputation and relational rewards that come with it. Being a sustainable company doesn't guarantee new business, but like good landscaping and home improvements, it definitely helps your "resale value." With the growing number of consumers who are driven by environmental consciousness when making buying decisions or showing brand loyalty, not being socially and environmentally responsible will no doubt count against you. By leading the charge for your company to go solar, you'll actually be serving as a public relations and corporate reputation genius.

GENERATIONAL IMPACT

This dovetails nicely with the next chapter—where I'll explain how solar is the right choice for today and how each of us (especially millennials) can leave a legacy through solar advocacy. Without stealing my own thunder, I encourage you to see the bigger picture of being part of the solution. Being your company's solar champion has immediate and long-term rewards. The ultimate impact is bigger than all of us living and working in the world today.

SOLAR FACTOID

The Solar Energy Industry Association reports that a new solar customer is currently being signed every three minutes in the United States.

CHAPTER EIGHT

A BETTER WORLD

THE BEST LEGACY? BEING PART OF THE SOLUTION

Relevance. It's what today's society demands more than ever before. It's what businesses and organizations struggle to achieve. It's what we all hope to achieve when we reach the far side of our life's journey. Will our lives have mattered when it's all said and done? What legacy will we leave for our children and their grand-children? There are so many ways we can each make a difference and leave the world a better place than when we arrived.

It's no secret that I believe solar energy and microgrid technology will play a significant and relevant role in changing the world as we know it, shifting the economic and political land-scapes to the advantage of the individual.

Think of all the progress we've seen over the last decade. For those of us who have been around for a few of them, the changes took place right under our noses while we were busy trying to be relevant. That progress has grown up alongside millennials, and Generation Z may never fully realize how big the world once was.

When we think of well-known brands like Apple, Droid, Facebook, Twitter, Pinterest, Instagram, and a growing multitude of digital and technological solutions, a quick glance over our shoulder into the not-so-distant past will show us a very different world than the one of iOS, *likes*, and hashtags we live in now.

We even consume media differently than we did before. Not long ago we watched the Super Bowl on one channel. Now it's also broadcast on a separate high-definition (HD) channel, on Spanish networks, or streamed online. And it can offer behind-the-scenes and special features to download, interact with, and share, all on multiple electronic devices while watching the game. And if a commercial isn't funny or interesting enough, we can always fast-forward or skip it entirely.

We've seen the rise of healthier lifestyles and the decline of fast food. We still drink, but we do it more responsibly. Texting while driving has become a leading cause of auto accidents. We buy lattes with our smartphones and pay our taxes with plug-and-play software sold to the masses. Even our watches are more functional computers than the ones we relied on in the past. We control what we watch and listen to; what we like, share, block, and report; and what we buy and sell. We can do all this whenever we want and from virtually anywhere. And just look at how well we've adapted. It hasn't been without hiccups, false starts, and do-overs, but all except the most bullheaded of us have engaged and embraced progress in one form or another.

Look ahead a decade into the future. Flying cars may still be a long way off, but it's not unlikely that zero-emission vehicles will become increasingly accessible to and adopted by the masses. We may see commercial airlines heading that way before long as well, given the fact that solar-powered airplanes are now circling the globe. The possibilities for technological progress are limited only by our imaginations and ingenuity. But imagine a world where only the sun, water, and air were relied on for power. Imagine what it will be like when each citizen generates and sustains his or her own energy supply and consumption.

Imagine the world that your children's children could grow up in—where the rising price of gas is no longer a problem because petroleum no longer powers vehicles and where rolling blackouts, planned shortages, and regularly scheduled rate increases for electricity are no longer talked about, because microgrids free everyone from grid dependence.

We are witnessing the dawn of that future right now. I'm not talking about an unlikely utopian society where personality and personal choice are removed for the greater good, and we all move slowly about wearing togas and thinking nice thoughts. I'm talking about a world created by people like you and me—who want to make wiser choices, who understand and exercise freedom of personal choice every day, and who are unafraid to pursue a better way that leads to a better life and ensures that we leave a positive legacy for future generations of our families.

As I shared earlier, solar as a primary source of power is already a reality in other countries, where homes and hospitals have no access to power grids and the rising cost of petroleum fuel for generators hits hardest of all. It's worth asking ourselves—if people who have so much less than what we are blessed with in this country are able to appreciate and gratefully accept solar energy as a lifesaving and sustainable resource, why are we so reluctant to do the same? Another question might be: Why do our government organizations support solar as a primary resource for under-

developed countries while resisting the growth of privately owned and controlled solar at home? And an even better question may be: If we had no grid or fossil fuels to rely on, wouldn't we just as quickly embrace solar as other countries already do? At some point, this question will no longer be hypothetical as the cost to access fossil fuels rises exponentially and as the cost of solar plus storage plunges.

Solar power and microgrids are already being installed as part of new construction from the ground up in countries that don't enjoy access to or the convenience of centralized monopolies spoon-feeding them their power. The wealth that still makes this country a land of opportunity is also one of our greatest weaknesses. Even the poorest among us have far more or access to it than those living in real poverty around the world. Our wealth makes it easier to take what we have for granted, to put off what we don't choose to do today, and to rely entirely on a false sense of security that tells us we will never run out.

Have no doubt, the microgrid will soon allow us to power our homes, charge our vehicles, and live free from grid reliance in America, but as with all progress, it will take a village to make it happen. That village is made up of individuals of "we." We have the choice, the freedom, and the technology. Affordability and accessibility are already here if we choose to take advantage of them. Eventually, the powers that be will be forced to sell off their interests in fossil fuel and go all in with alternative energy. When they finally do, the amount of choice, control, and personal savings will be in their hands.

Entire communities are utilizing microgrids and cutting the cord with utilities. Not out of necessity because of location but by choice. According to Wikipedia, Wildpoldsried in Germany began a series of projects to produce renewable energy, including wind turbines and biomass digesters for cogeneration of heat and power. Since then, new work has included a number of energy conservation projects, more wind and biomass use, small hydro plants, photovoltaic panels on private houses, and district heating, as well as ecological flood control and wastewater systems.

Today, nine new community buildings, including a school, gymnasium, and community hall, complete with solar panels, have been constructed as a result of this unforeseen level of prosperity. There are three companies operating four biogas digesters with a fifth under construction and seven windmills with two more on the way. Almost 200 private households are equipped with solar, which pays them dividends. The district heating network has 42 connections, and there are three small hydro power plants. "By 2011, Wildpoldsried produced 321 percent more energy than it needs and generated 4.0 million Euro in annual revenue. At the same time, there was a 65 percent reduction in the town's carbon footprint."

I can't urge you enough. Don't allow indifference, ignorance, or codependency to make you an unwitting part of the problem. Be a critical thinker. Do the research. Make independence a priority, and be part of the solution. Do that, and your life will have relevance for generations to come.

SOLAR FACTOID

It is estimated that Americans pay more than $150 billion a year for costs associated with power outages on the grid.

CHAPTER NINE

A NEW ERA

EMPOWERED TO BE INDEPENDENT

Sunsets are magnificent, but I especially love to watch the sunrise because it means I'm breathing, I'm alive another day, and the dawn signals another opportunity to be relevant, to be daring, to be part of the solution. And to me, it means another day to harvest electricity.

In the context of solar, almost all Americans still have a daily opportunity to not only appreciate the beauty of the rising sun but to harness its power for their own benefit. Around 1 percent of our country is currently using solar, and with this book and our efforts daily, we're doing all we can to change the paradigms of the other 99 percent.

As we've discovered, the sun has been shining high and bright as a trusted source of power far longer than we've been dependent on finite and hotly contested crude resources. And just as solar power has transformed lives by making the few using it in our country less dependent on the grid, the microgrid will take it to another level.

Lithium-ion batteries were not always around (let alone all of the electronics we rely on that they power) and yet, as with all progress, now we buy them in pairs or in bricks today because they're small enough, can be mass-produced, and are affordable enough to make them convenient as an accepted necessity of daily American life. In addition, they future proof your investment in storage by virtue of being a modular portion of your system, which means you can add or replace batteries as technology advances even more.

If we think back, the computer's story is not all that different. In the beginning, they were massive in size and expense and limited in their capacity—while simultaneously holding enormous potential as well. Then came the personal computer that only a select few could own. They were no longer massive in size, but they were still cost prohibitive for most, and they had only begun to scratch the surface in what they would ultimately do. Today, we wear them on our wrists, use them as our phones, and use them to write languages (codes) that have and continue to change the world.

If I told you that around 1 percent of Americans are currently using computers, you'd close this book and write me off—and rightly so. The reason microgrids are only just now on the verge of breakthrough to the masses is for similar challenges with size, cost, and capacity. Even a large portion of the 1 percent who have smartly chosen solar are using it without battery backup for those very reasons.

The massive cubes that weigh several tons and are being used to power small villages, military bases, historic attractions such as Alcatraz Island, and even college campuses, were not at all practical for Mr. and Mrs. Jones—with cars in the garage, pools in the yard, and no other available space to speak of in suburbia. And most of the folks living in rural areas, who may have the acreage, might not have had the discretionary income to finance one of these micro-beasts when the cost was deemed too high for even affluent, private solar-system owners.

But like computers before them, the batteries that (with a PV solar system) create the closed-loop microgrid system have evolved even beyond what you have seen already. I know because Mycrogrid and our advanced energy storage partners have developed and are rolling them out for mass distribution.

Currently, batteries for residential and business owners are at or around the original personal computer stage. They're small enough to house, still pricey, and have limited capacity but are full of massive potential. It's that potential that we have focused on in our development of Mycrogrid. This is a topic we'll dive into in the last chapter.

Even with progress, there's not a battery that's small enough for mass adoption with the capacity to store enough energy to power an entire home indefinitely, let alone the neighborhood and beyond. With rapid advances in technology, it's coming soon—but currently microgrids generally excel at three things:

1. Providing energy backup in brownouts, blackouts, and natural disasters.

2. Providing enough storage to power the largest energy hogs in the home or office. They are charged with solar by day and allow the owner to draw smaller amounts of power from the grid at night when demand on the system and rates are lower—energy arbitrage.

3. Discharge your batteries to eliminate demand charges.

Going back to the lithium-ion battery comparison, microgrid storage batteries are similar in that they will become a convenient necessity over time. But in regard to supplying power, the copper tops you snap into your electronics are actually more like the grid because they have a finite capacity and shelf life. You have to keep paying for them to benefit from them, you have to rely on the distributor to make them available to you, and they don't always work as advertised.

Since microgrid storage batteries are rechargeable using lithium-ion chemistry in them, there is typically more up-front cost than prior generations of rechargeable batteries that used lead-acid or similar chemistry. However, lithium-ion batteries are designed to last for much longer periods of time. They also have a greater energy density, allowing for more storage in the same footprint and adding even more to their value!

If a solar system has the ability to roll back your meter, then how much better is it that instead of storing that extra energy on the grid at the expense of being tied to the utility company, you're able to bank it in storage savings and hold onto *your* energy?

Not only has the engineering, design, and manufacturing of microgrid battery storage progressed, so has its technology. Gone are the days of dated control units that were as intuitive for some as programming an irrigation control box. Today, these systems are controlled with the same beautiful, intuitive, digital interface we've grown accustomed to with our personal devices, allowing you to touch a screen and instantly change your home or business' electric profile with the energy management certainty provided by your microgrid.

In the future that you and I can bring about together, not only solar but also wind, geothermal, and who knows what other renewable energy technologies, will provide all power necessary for the closed-loop microgrid, setting the owner free from the unreliable and costly grid forever. Think about what that will mean. That's unprecedented security, my friend! It's easy to miss it if we stop at "no more worrying about power outages." If it was that superficial, the benefit may only mean fewer candles burned or shins bruised.

What does this mean?

✓ no more replacing all of those spoiled groceries

✓ no more missing the final shot of the big game or the "ah-ha" moment of your favorite movie when your television goes dark

✓ no more cold showers at Christmas when the electric water heater goes out

✓ no more being vulnerable when some utility worker you've never met, working in a place you've never been to, takes out the grid by accidentally clipping a power line

Speaking of vulnerable, what about those natural disasters? Here in California, it's earthquakes and fire. Elsewhere there are tornadoes, hurricane seasons, nor'easters, tsunamis, mudslides, and floods. There are few places in our great land where none of the above are a risk, and if we can't control them, you can bet your electric bill that the utilities have no control over them either. Freedom from the risk of the power outages that inevitably occur as a result should be enough to inspire anyone to at least consider the solar and microgrid alternative sooner than later.

When combined with the renewable and independent energy supply of solar, these batteries may only be a life raft as of the writing today in the summer of 2015, but the microgrid closed-loop system that's dawning will be a mighty vessel that can withstand the fiercest storms while the grid sinks like the Titanic that it is.

Let's review:

- ✓ In chapter 1, we debunked the stereotypes associated with living off the grid.

- ✓ In chapter 2, we busted commonly held myths surrounding solar.

- ✓ In chapter 3, we—hopefully—pushed your hot buttons for going solar and getting ready to add battery storage.

- ✓ In chapter 4, we showed you how solar will help you keep more of your hard-earned money.

- ✓ In chapter 5, we showed how you can help save the environment while you save money.

- ✓ In chapter 6, we showed how being a part of this solar renaissance serves to overcome generational stereotypes.

- ✓ In chapter 7, we touched on the benefits of implementing solar and microgrid technology for savvy business owners and the brilliant advocates who plant the seeds in the corporate paradigm.

✓ In chapter 8, we talked about relevance, legacy, and how what we do to bring in the microgrid era today will impact all lives of future generations.

✓ And we've just taken a final look at what that world will look like as we witness the dawning of that era even now.

After all of that, my sincere hope is that you have a better appreciation, understanding, passion, and motivation to embrace solar, to be a microgrid adopter at home and the workplace, and to take action now.

Microgrids are powered by solar, wind, geothermal, and other renewable means that haven't even been developed yet. Imagine it. No more rolling blackouts. No more power outages in bad weather. No reliance on an unstable grid or vulnerability to human errors that leave you inconvenienced, losing money on lost perishables, and losing time to lost work and data when your computer goes down.

The only thing that makes all of that microgrid-generated freedom a possibility for tomorrow, rather than a reality today, is the battery. There are many, many great minds working on battery technology as I've pointed out, but being the type of competitor who wants to be in the game and not watching from the sidelines, I decided to jump into the fray and have been working with many of the aforementioned great minds on a way to deliver greater microgrid capability to you.

CASE STUDY 3

The young couple who purchased "Indio Two", so named because of their city, had Mycrogrid install storage in their home that features a 16 kWh Sonnenbatterie system paired with their existing 12 kW DC SunPower solar system. Because

their primary focus for storage was energy management and the ability to save money, they wanted to be strategic about when the batteries were discharged in order to help eliminate any peak charges or energy costs. They were most excited about Mycrogrid's monitoring, which gave them a second-by-second, granular view of their true solar output, their home's energy consumption, and battery state-of-charge 24-7 on their mobile phones. All updates to this web-based monitoring were done automatically.

As battery storage density increases in the short run, the Mycrogrid staff can exchange their lithium-ion batteries for the most advanced and compact storage batteries. Mycrogrid will save them money, and at the same time, optimize their utility's central electric grid. Their system, along with the tens of thousands that will come online in the next few years, will reduce peak energy demand while contributing to the reliability of the grid, which means deferring expensive grid maintenance updates that cause additional cost to utility ratepayers.

CHAPTER TEN

MYCROGRID®

Quoting myself in full-on solar geek mode, "Understanding the future of distributed natural energy generation (microgrids) lies in storage solutions (batteries). In 2014, I founded Mycrogrid to offer advanced smart-battery systems to augment energy savings in conjunction with leading battery manufacturers." We went with Mycrogrid for exactly how it reads—it's *my* microgrid ... my power, my choice.

Mycrogrid delivers energy cost savings from:

- ✓ Reduced purchases of grid power and transmission services

- ✓ Reduced resource interconnection cost

- ✓ Reliability and power quality improvements for the existing, aging macrogrid

- ✓ Greater integration of renewables—reduced carbon and other emissions

- ✓ Greater security and safety during macrogrid outages—safe haven

And of course, Mycrogrid was created in hopes that the increased accessibility and more intuitive and capable system would encourage greater participation from folks like you to expedite grid modernization and move as many as we could toward a decentralized grid architecture and a reduced reliance on utility scale resources. The current Mycrogrid system we use for our emergency backup and to mitigate energy usage for our major office/kitchen appliances will fit in your garage and is monitored and controlled from a digital touch pad screen.

Mycrogrid is simply the march of technology-enabled independence expanding to the realm of renewable energy. You can control your house lights and security on your smartphone from another country. Before long, you'll be able to do the same with your personal microgrid using a battery no bigger than a paper shredder and with no need for a meter.

As America's first NABCEP-accredited solar installation company (North American Board of Certified Energy Practitioners), Renova Solar services residential and commercial clients in Southern California. After all, we said you have to eat the elephant one bite at a time—California is an elephant in its own right. But if you are one of our Southern California neighbors, you haven't gone solar yet, and my words have inspired you to take a closer look, please reach out. We'll be honored to help you make an informed solar decision—whether or not we provide you with your new system.

As for our Mycrogrid battery storage offering, we have every intention of making this technology available to any and all who want to be a part of the cord-cutting renaissance—no matter where you live in the United States.

Which leads me to my formal invitation …

FINAL THOUGHTS

At Renova Solar we're known as the "Local. Brighter. Better." choice for solar design, installations, and maintenance on all solar systems in the area (ours or theirs) through our RenovaPLUS inspection and maintenance program (solar systems need regular care like anything else of great value). With our Mycrogrid solutions, we're earning the right to be known nationally as a leader in closed-loop, microgrid technology.

Now you know how it works (solar + battery = microgrid).

You know that the technology is already here and ready to use.

You know how microgrids deliver independence, reliability, savings, and security.

You know how Mycrogrid systems are more cost-effective than solar alone.

And hopefully, you now share my vision for the brighter tomorrow that's just around the corner.

The only thing left is for you to take ownership of the decision yourself. If you want to learn more about solar and whether or not it's right for you, I invite you to visit our website at www.renova. solar today. If you're ready to delve more into microgrids, find that at www.mycrogrid.solar.

I offer you my sincerest thanks for investing your time with me here, and I hope that you consider yourself better informed for having read this.

GLOSSARY

As a resource, we've added this solar glossary with definitions for some of the most common technical terms related to solar power and photovoltaic (PV) technologies, including those having to do with electricity, power generation, and concentrating solar power (CSP) from the Office of Energy Efficiency & Renewable Energy. More definitions can be found at energy.gov/eere/sunshot/solar-energy-glossary.

AES.* An acronym for Advanced Energy Storage—technologies that convert electricity into energy, store it, and then convert it back into usable electricity at a later time.

alternating current (AC). A type of electrical current, the direction of which is reversed at regular intervals or cycles. In the United States, the standard is 120 reversals or 60 cycles per second. Electricity transmission networks use AC because voltage can be controlled with relative ease.

annual solar savings. The annual solar savings of a solar building is the energy savings attributable to a solar feature relative to the energy requirements of a non-solar building.

array. See ***photovoltaic (PV) array***.

array current. The electrical current produced by a photovoltaic array when it is exposed to sunlight.

array operating voltage. The voltage produced by a photovoltaic array when exposed to sunlight and connected to a load.

autonomous system. See *stand-alone system.*

balance of system. Represents all components and costs other than the photovoltaic modules/array. It includes design costs, land, site preparation, system installation, support structures, power conditioning, operation and maintenance costs, indirect storage, and related costs.

battery. Two or more electrochemical cells enclosed in a container and electrically interconnected in an appropriate series/parallel arrangement to provide the required operating voltage and current levels. Under common usage, the term battery also applies to a single cell if it constitutes the entire electrochemical storage system.

battery available capacity. The total maximum charge, expressed in ampere-hours, that can be withdrawn from a cell or battery under a specific set of operating conditions, including discharge rate, temperature, initial state of charge, age, and cut-off voltage.

battery capacity. The maximum total electrical charge, expressed in ampere-hours, which a battery can deliver to a load under a specific set of conditions.

battery cell. The simplest operating unit in a storage battery. It consists of one or more positive electrodes or plates, an electrolyte that permits ionic conduction, one or more negative electrodes or plates, separators between plates of opposite polarity, and a container for all the above.

battery cycle life. The number of cycles, to a specified depth of discharge, that a cell or battery can undergo before failing to meet its specified capacity or efficiency performance criteria.

battery energy capacity. The total energy available, expressed in watt-hours (kilowatt-hours), which can be withdrawn from a fully charged cell or battery. The energy capacity of a given cell varies with temperature, rate, age, and cut-off voltage. This term is more common to system designers than it is to the battery industry where capacity usually refers to ampere-hours.

battery energy storage. Energy storage using electrochemical batteries. The three main applications for battery energy storage systems include spinning reserve at generating stations, load leveling at substations, and peak shaving on the customer side of the meter.

battery life. The period during which a cell or battery is capable of operating above a specified capacity or efficiency performance level. Life may be measured in cycles and/or years, depending on the type of service for which the cell or battery is intended.

California Solar Energy Industries Association (CALSEIA). * A nonprofit which is focused on advancing the common interests of the California solar industry, helping make California's solar market the most robust in the United States.

capacity (C). See *battery capacity.*

cell (battery). A single unit of an electrochemical device capable of producing direct voltage by converting chemical energy into electrical energy. A battery usually consists of several cells electrically connected together to produce higher voltages (Sometimes the terms cell and battery are used interchangeably). See also *photovoltaic (PV) cell.*

charge. The process of adding electrical energy to a battery.

concentrating photovoltaics (CPV). A solar technology that uses lenses or mirrors to concentrate sunlight onto high-efficiency solar cells.

concentrating solar power (CSP). A solar technology that uses mirrors to reflect and concentrate sunlight onto receivers that convert solar energy to heat. This thermal energy is then used to produce electricity with a steam turbine or heat engine driving a generator.

concentrator. A photovoltaic module, which includes optical components such as lenses (Fresnel lens) to direct and concentrate sunlight onto a solar cell of smaller area. Most concentrator arrays must directly face or track the sun. They can increase the power flux of sunlight hundreds of times.

conductor. The material through which electricity is transmitted, such as an electrical wire or transmission or distribution line.

conversion efficiency. See *photovoltaic (conversion) efficiency.*

current. See *electric current.*

cycle. The discharge and subsequent charge of a battery.

days of storage. The number of consecutive days the stand-alone system will meet a defined load without solar energy input. This term is related to system availability.

direct current (DC). A type of electricity transmission and distribution by which electricity flows in one direction through the conductor, usually relatively low voltage and high current, to be used for typical 120-volt or 220-volt household appliances. DC must be converted to alternating current, its opposite.

discharge. The withdrawal of electrical energy from a battery.

discharge factor. A number equivalent to the time in hours during which a battery is discharged at constant current, usually expressed as a percentage of the total battery capacity, i.e., C/5 indicates a discharge factor of five hours. Related to *discharge rate.*

discharge rate. The rate, usually expressed in amperes or time, at which electrical current is taken from the battery.

disconnect. Switch gear used to connect or disconnect components in a photovoltaic system.

distributed energy resources (DER). A variety of small, modular power-generating technologies that can be combined with energy management and storage systems and used to improve the operation of the electricity delivery system, whether or not those technologies are connected to an electricity grid.

distributed generation. A popular term for localized or on-site power generation.

distributed power. Generic term for any power supply located near the point where the power is used opposite of central power. See also *stand-alone systems.*

distributed systems. Systems that are installed at or near the location where the electricity is used, as opposed to central systems that supply electricity to grids. A residential photovoltaic system is a distributed system.

electric current. The flow of electrical energy (electricity) in a conductor, measured in amperes.

electrical grid. An integrated system of electricity distribution, usually covering a large area.

electricity. Energy resulting from the flow of charge particles, such as electrons or ions.

electron. An elementary particle of an atom with a negative electrical charge and a mass of 1/837 of a proton; electrons surround the positively charged nucleus of an atom and determine the chemical properties of an atom. The movement of electrons in an electrical conductor constitutes an electric current.

energy. The capability of doing work; different forms of energy can be converted to other forms, but the total amount of energy remains the same.

energy asset arbitrage. * The practice of taking advantage of a price difference between two or more items to strike a combination of matching deals that capitalizes on the imbalance.

energy audit. A survey that shows how much energy is used in a home, which helps find ways to use less energy.

energy density. The ratio of available energy per pound, usually used to compare storage batteries.

feed-in tariff (FIT). A renewable energy policy that typically offers a guarantee of: payments to project owners for total kWh of renewable electricity produced; access to the grid; and stable, long-term contracts (15-20 years). Also called fixed-price policies, minimum price policies, standard offer contracts, feed laws, renewable energy payments, renewable energy dividends, and advanced renewable tariffs.

future proof. * To design a system that is modular in nature so that individual components can be added and/or replaced as needed when more components are needed or more advanced components become available, as in the replacement of batteries in a microgrid system as newer models which store more energy are developed.

grid. See *electrical grid.*

grid-connected system. A solar electric or photovoltaic (PV) system in which the PV array acts like a central generating plant, supplying power to the grid.

grid-interactive system. See *grid-connected system.*

inverter. A device that converts direct current electricity to alternating current either for stand-alone systems or to supply power to an electricity grid.

IOU. * Investor-owned utility.

kilowatt (kW). A standard unit of electrical power equal to 1,000 watts, or to the energy consumption at a rate of 1,000 joules per second.

kilowatt-hour (kWh). 1,000 watts acting over a period of one hour. The kWh is a unit of energy 1 kWh=3600 kJ.

life. The period during which a system is capable of operating above a specified performance level.

life-cycle cost. The estimated cost of owning and operating a photovoltaic system for the period of its useful life.

light-induced defects. Defects, such as dangling bonds, induced in an amorphous silicon semiconductor upon initial exposure to light.

load. The demand on an energy-producing system; the energy consumption or requirement of a piece or group of equipment is usually expressed in terms of amperes or watts in reference to electricity.

load current (A). The current required by the electrical device.

macrogrid. *Another word for the grid used by a utility.

megawatt (MW). 1,000 kilowatts or 1 million watts; standard measure of electric power plant generating capacity.

megawatt-hour. 1,000 kilowatt-hours or 1 million watt-hours.

microgrid. * A system which is comprised of a renewable energy source such as solar coupled with advanced energy storage, most often batteries.

module. See *photovoltaic (PV) module.*

Mycrogrid/RenovaPLUS Mycrogrid. *A company launched specifically to design, engineer, manufacture, and install components related to microgrid energy systems.

National Electrical Code (NEC). Contains guidelines for all types of electrical installations. The 1984 and later editions of the NEC contain Article 690, "Solar Photovoltaic Systems," which should be followed when installing a PV system.

North American Board of Certified Energy Practitioners (NABCEP). *A nonprofit corporation with a mission to support and work with the renewable energy and energy efficiency industries, professionals, and stakeholders to develop and implement quality credentialing and certification programs for practitioners.

overcharge. Forcing current into a fully charged battery. The battery will be damaged if overcharged for a long period.

panel. See *photovoltaic (PV) panel.*

peak demand/load. The maximum energy demand or load in a specified time period.

photon. A particle of light that acts as an individual unit of energy.

photovoltaic(s) (PV). Pertaining to the direct conversion of light into electricity.

photovoltaic (PV) array. An interconnected system of PV modules that function as a single electricity-producing unit. The modules are assembled as a discrete structure, with common support or mounting. In smaller systems, an array can consist of a single module.

photovoltaic (PV) cell. The smallest semiconductor element within a PV module to perform the immediate conversion of light into electrical energy (direct current voltage and current). Also called a **solar cell.**

photovoltaic (PV) conversion efficiency. The ratio of the electric power produced by a photovoltaic device to the power of the sunlight incident on the device.

photovoltaic (PV) device. A solid-state electrical device that converts light directly into direct current electricity of voltage-current characteristics that are a function of the characteristics of the light source and the materials in and design of the device. Solar photovoltaic devices are made of various semiconductor materials, including silicon, cadmium sulfide, cadmium telluride, and gallium arsenide, and in single crystalline, multicrystalline, or amorphous forms.

photovoltaic (PV) effect. The phenomenon that occurs when photons, the "particles" in a beam of light, knock electrons loose from the atoms they strike. When this property of light is combined with the properties of semiconductors, electrons flow in one direction across a junction, setting up a voltage. With the addition of circuitry, current will flow and electric power will be available.

photovoltaic (PV) generator. The total of all PV strings of a PV power supply system, which are electrically interconnected.

photovoltaic (PV) module. The smallest environmentally protected, essentially planar assembly of solar cells and ancillary parts, such as interconnections, terminals, and protective devices such as diodes intended to generate direct current power under unconcentrated sunlight. The structural (load carrying) member of a module can either be the top layer (superstrate) or the back layer (substrate).

photovoltaic (PV) panel. Often used interchangeably with PV module (especially in one-module systems), but more accurately used to refer to a physically connected collection of modules (i.e., a laminate string of modules used to achieve a required voltage and current).

photovoltaic (PV) system. A complete set of components for converting sunlight into electricity by the photovoltaic process, including the array and balance of system components.

photovoltaic-thermal (PV/T) system. A photovoltaic system that, in addition to converting sunlight into electricity, collects the residual heat energy and delivers both heat and electricity in usable form. Also called a total energy system or solar thermal system.

power. The amount of electrical energy available for doing work, measured in horsepower, watts, or Btu per hour.

power conditioning. The process of modifying the characteristics of electrical power (e.g., inverting direct current to alternating current).

power conditioning equipment. Electrical equipment, or power electronics, used to convert power from a photovoltaic array into a form suitable for subsequent use. A collective term for inverter, converter, battery charge regulator, and blocking diode.

power conversion efficiency. The ratio of output power to input power of the inverter.

power density. The ratio of the power available from a battery to its mass (W/kg) or volume (W/l).

PV. See *photovoltaic(s).*

remote systems. See *stand-alone systems.*

shelf life of batteries. The length of time, under specified conditions, that a battery can be stored so that it keeps its guaranteed capacity.

silicon (Si). A semi-metallic chemical element that makes an excellent semiconductor material for photovoltaic devices. It crystallizes in face-centered cubic lattice like a diamond. It's commonly found in sand and quartz (as the oxide).

smart grid. An intelligent electric power system that regulates the two-way flow of electricity and information between power plants and consumers to control grid activity.

soft costs. Non-hardware costs related to PV systems, such as financing, permitting, installation, interconnection, and inspection.

solar cell. See **photovoltaic (PV) cell.**

solar energy. Electromagnetic energy transmitted from the sun (solar radiation). The amount that reaches the earth is equal to one billionth of total solar energy generated or the equivalent of about 420 trillion kilowatt-hours.

Solar Energy Industries Association (SEIA).* A United States nonprofit member organization whose mission is to build a strong solar industry through advocacy and education.

solar panel. See **photovoltaic (PV) panel.**

solar thermal electric systems. Solar energy conversion technologies that convert solar energy to electricity by heating a working fluid to power a turbine that drives a generator. Examples of these systems include central receiver systems, parabolic dish, and solar trough.

stand-alone system. An autonomous or hybrid photovoltaic system not connected to a grid. May or may not have storage, but most stand-alone systems require batteries or some other form of storage.

storage battery. A device capable of transforming energy from electric to chemical form and vice versa. The reactions are almost completely reversible. During discharge, chemical energy is converted to electric energy and is consumed in an external circuit or apparatus.

string. A number of photovoltaic modules or panels interconnected electrically in series to produce the operating voltage required by the load.

system storage. See **battery capacity.**

watt. The rate of energy transfer equivalent to one ampere under an electrical pressure of one volt. One watt equals 1/746 horsepower or one joule per second. It is the product of voltage and current (amperage).

*Definitions have been added by the author.

BIBLIOGRAPHY

Battaglia, Alessandro. "On the Methods and Convenience of Using Solar Heat for Steam Engines." Presented at the Encouragement Institute, Naples, Italy, 1884.

Berkley Lab, "Berkeley Lab Illuminates Price Premiums for U.S. Solar Home Sales," news release, January 13, 2015, https://newscenter.lbl.gov/2015/01/13/berkeley-lab-illuminates-price-premiums-u-s-solar-home-sales/.

Greg Rouse. and John Kelly, "Electricity Reliability: Problems, Progress and Policy Solutions," Galvin Electricity Initiative, 2011, http://www.galvinpower.org/sites/default/files/Electricity_Reliability_031611.pdf, accessed June 1, 2015.

"Hurricane Sandy: Covering the Storm," *New York Times*, November 6, 2012, http://www.nytimes.com/interactive/2012/10/28/nyregion/hurricane-sandy.html, accessed June 1, 2015.

Mouchot, Augustine. *La Chaleur solaire et ses applications industrielles*. Gauthier-Villars: Paris, 1869.

Net Metering, *Issues & Policies*, SEIA website, http://www.seia.org/policy/distributed-solar/net-metering, accessed June 1, 2015.

"Solar Industry Data," Research & Resources, SEIA website, http://www.seia.org/research-resources/solar-industry-data, accessed June 1, 2015.

Stephen Lacey, "Verizon Executive on $40M Solar Investment: 'It's About Driving Shareholder Value,'" August 26, 2014, Greentech Media website, http://www.greentechmedia.com/articles/read/verizon-executive-on-investing-in-solar-its-about-driving-shareholder-value, accessed June 1, 2015.

BP Reaches $18.7 billion settlement over deadly 2010 spill. Wade, T., Reuters, July 2, 2015.

http://link.reuters.com/duz94w

CPSIA information can be obtained at www.ICGtesting.com
Printed in the USA
BVOW04s0931211015

423490BV00021B/136/P